How Well Do Executives Trust Their Intuition?

Data Analytics Applications

Series Editor: Jay Liebowitz

PUBLISHED

Actionable Intelligence for Healthcare
by Jay Liebowitz and Amanda Dawson
ISBN: 978-1-4987-6665-4

Analytics and Knowledge Management
by Suliman Hawamdeh and Hsia-Ching Chang
ISBN 978-1-1386-3026-0

Big Data Analytics in Cybersecurity
by Onur Savas and Julia Deng
ISBN: 978-1-4987-7212-9

Big Data and Analytics Applications in Government
Current Practices and Future Opportunities
by Gregory Richards
ISBN: 978-1-4987-6434-6

Big Data in the Arts and Humanities
Theory and Practice
by Giovanni Schiuma and Daniela Carlucci
ISBN 978-1-4987-6585-5

Data Analytics Applications in Education
by Jan Vanthienen and Kristoff De Witte
ISBN: 978-1-4987-6927-3

Data Analytics Applications in Latin America and Emerging Economies
by Eduardo Rodriguez
ISBN: 978-1-4987-6276-2

Data Analytics for Smart Cities
by Amir Alavi and William G. Buttlar
ISBN 978-1-138-30877-0

Data-Driven Law
Data Analytics and the New Legal Services
by Edward J. Walters
ISBN 978-1-4987-6665-4

Intuition, Trust, and Analytics
by Jay Liebowitz, Joanna Paliszkiewicz, and Jerzy Gołuchowski
ISBN: 978-1-138-71912-5

Research Analytics
Boosting University Productivity and Competitiveness through Scientometrics
by Francisco J. Cantú-Ortiz
ISBN: 978-1-4987-6126-0

Sport Business Analytics
Using Data to Increase Revenue and Improve Operational Efficiency
by C. Keith Harrison and Scott Bukstein
ISBN: 978-1-4987-8542-6

How Well Do Executives Trust Their Intuition?

Edited by
Jay Liebowitz, Yolande Chan, Tracy Jenkin,
Dylan Spicker, Joanna Paliszkiewicz
and Fabio Babiloni

CRC Press
Taylor & Francis Group
Boca Raton London New York

CRC Press is an imprint of the
Taylor & Francis Group, an **Informa** business
AN AUERBACH BOOK

CRC Press
Taylor & Francis Group
6000 Broken Sound Parkway NW, Suite 300
Boca Raton, FL 33487-2742

First issued in paperback 2022

© 2019 by Taylor & Francis Group, LLC
CRC Press is an imprint of Taylor & Francis Group, an Informa business

No claim to original U.S. Government works

ISBN 13: 978-1-03-247582-0 (pbk)
ISBN 13: 978-1-138-49262-2 (hbk)
ISBN 13: 978-0-429-02226-5 (ebk)

DOI: 10.1201/9780429022265

**Visit the Taylor & Francis Web site at
http://www.taylorandfrancis.com**

**and the CRC Press Web site at
http://www.crcpress.com**

Dedication

I would like to dedicate this book to John Wyzalek, Rich O'Hanley, and the other senior executives at Taylor & Francis who had the foresight and intuitive awareness to publish this book and the entire Data Analytics Applications book series. (Jay)

I dedicate this book to all executives whose hearts and minds indicate very different courses of action. May this book help them bridge the painful divide and take the best ways forward. (Yolande)

To those who are in search of the forest among the trees but who are wise enough to bring a map just in case. (Tracy)

Kennedy, without you none of my work would be possible. Thank you. (Dylan)

This is for you, Natalie. Thanks for being a wonderful daughter and making my world delightful. You are the joy and pride of my life. (Joanna)

"We are such stuff as dreams are made on."
(Shakespeare, The Tempest) (Fabio)

Contents

Contributors

Susan Aldridge
Senior Vice President, Online
   Learning, Drexel University
President, Drexel University
   Online
Philadelphia, Pennsylvania

Fabio Babiloni
Professor of Biomedical
   Engineering
University of Rome Sapienza
Rome, Italy

Yolande Chan
E. Marie Shantz Professor of IT
   Management and Associate
   Dean of Research
Smith School of Business
Queen's University
Kingston, Ontario, Canada

Patrizia Cherubino
University of Rome Sapienza
Rome, Italy

Chiara Corso
Deloitte
Rosslyn, Virginia

David Dye
Director, Federal Human Capital
   Services
Deloitte
Rosslyn, Virginia

Holly Morgan Frye
Assistant Vice President, Student
   Affairs
Director, Community Outreach
Shepherd University
Shepherdstown, West Virginia

Ralph Tilden (RT) Good
Dean and Professor, College of
   Business and Management
Lynn University
Boca Raton, Florida

Eugene W. Grant
Mayor
Seat Pleasant, Maryland

William Hall
Founding Member, Fratelli
   Bologna
Co-Founder, BATS Improv
San Francisco, California

Tracy A. Jenkin
Associate Professor and
   Distinguished Faculty
   Fellow of MIS
   Smith School of Business
Queen's University
Kingston, Ontario, Canada

Claradith Landry
Deloitte
Rosslyn, Virginia

Jay Liebowitz
Distinguished Chair of Applied
    Business and Finance
Harrisburg University of Science
    and Technology
Harrisburg, Pennsylvania

Ben Martz
Dean, College of Business
Shepherd University
Shepherdstown, West Virginia

Enrica Modica
University of Rome Sapienza
Rome, Italy

Joanna Paliszkiewicz
Professor of WULS
Warsaw University of Life Sciences
Warsaw, Poland

Jennifer Rompre
Deloitte
Rosslyn, Virginia

Kyle Sandell
Deloitte
Rosslyn, Virginia

Dylan Spicker
University of Waterloo
Waterloo, Ontario, Canada

Rebecca Stockley
President Improv Lady
Co-Founder, BATS Improv
San Francisco, California

William Tanner
Deloitte
Rosslyn, Virginia

Part One

Intuition

1

Using Your Intuition

Written by: Jay Liebowitz

CONTENTS

Recent studies, including the KPMG Trust Gap Study, have shown that many CEOs rely more on their intuition for key decisions than strictly on their data analytics. Part of the reason is that the internal data quality in organizations is often spurious, causing a lack of confidence in any resulting data analytics from those sources.

Of course, applying "rational intuition" instead of becoming just a "hunch artist," as Peter Drucker used to say, may be a preferred approach in combining one's experiential learning with analysis.

In looking at how executives make decisions, there is usually a complementary set of both gut feeling and data-driven approaches. However, in this age of big data and analytics, the executive's intuitive feel and awareness are often minimized in favor of "going with the numbers." Intuitive awareness is real, and can also be extremely valuable, assuming that one has a good track record when benchmarking one's executive decisions against the results based on the intuitive decision.

One story highlighting the power of intuition relates to one of the Netflix senior executives who relied on his intuition, versus all the data produced beforehand, in making a decision whether to approve an initial show. Even though the data indicated that the show should not be approved, the executive went with his gut feeling, which signaled that it would be a great show. And, indeed, *House of Cards* was!

THE THREE I'S

When we look at intuition (Liebowitz et al., 2018), there is sometimes confusion between what we refer to as *the three I's*: intuition, insight, and instinct. In characterizing *intuition*, the following factors come to mind.

- Experience-driven
- Holistic
- Affective (closely connected with emotions; feelings-based signals)
- Quick (thinks quickly; primed for immediate action)
- Non-conscious (can't easily map logical audit trails, so to speak, to the judgment; knowing without knowing how one knows)

Simply put, instinct is typically innate, similar to a maternal instinct. Insight is awareness gleaned from deep understanding. Intuition is non-conscious reasoning. According to the Max Planck Institute for Human Development and research on gut decisions performed by Shabnam Mousavi at Johns Hopkins University, "It's clear that many business leaders believe there's a time to put data aside when thinking through a problem" (*Johns Hopkins Magazine*, Spring 2015). Mousavi cites research from 2012 that indicated "almost half of the managers (i.e., subscribers to Chief Executive magazine, CFO magazine, CFO Asia, and CFO Europe) consider their 'gut feel' an important or very important factor in making capital allocation decisions" (*Johns Hopkins Magazine*, Spring 2015). Mousavi's research examines the use of heuristics (rules of thumb acquired through experience) in executive decision making. However, well-known scientists such as Daniel Kahneman feel that heuristics may lead to errors and biases in judgments. Certainly, more research on, for example, *fast-and-frugal heuristics* (e.g., imitating the successful, etc.) is needed, but some of Mousavi's initial research shows that these types of heuristics can lead to more accurate judgments (*Johns Hopkins Magazine*, Spring 2015).

Data visualization can also support executive decision making. However, even in this case, research from Moore (2017) and others indicates that there must be an "intuitive" platform that enables collaboration and newness.

Even outside the business field, we see intuition being applied. For example, in medicine, we appreciate an evidence-based approach to diagnosing and treating patients. However, here again, intuition plays an interesting role, as shown by this quote from Scott Zeger, former provost at Johns Hopkins University (*Johns Hopkins Magazine*, Spring 2015):

> What we're looking to do is take that special brand of experience and *intuition* that the best doctors have and turn it into something that has a scientific foundation based in data and can eventually form useful, effective practice tools that we can put in the hands of clinicians here and around the country and the world.

There have been numerous studies over the years that have looked at the role of intuition in clinical nursing practice. The research results indicate that expert nurses apply their intuition accurately in most cases. Rovithis et al. (2015) indicate that recognition and acceptance of intuition as a valid part of nursing practice is recommended so that the nurses' confidence in their implementation of intuition will increase. Robert et al. (2014) found growing evidence suggesting that intuition in nursing is a critical part of effective decision making that supports safe patient care. Fox et al. (2016) also found that experienced clinicians are clustered into a single, common-factor response, which the researchers assert is due to the factor of intuition.

THE GOOD, THE BAD, AND THE UGLY

Intuition has been shown to be effective in time-sensitive and crisis situations. Marinos and Rosni's (2017) research at Lund University found that managers employed intuitive processes in volatile environments, information scarcity, and complex situations, and when an immediate decision was of the essence. They also concluded that managers favored intuition in strategic decision making in cases of high decision uncertainty and volatile environments. Furthermore, the University of Cambridge found that hedge fund traders who relied on their gut feelings outperformed those who didn't. Rauf (2014), in his study of 306 managers, found that the use of intuition is caused by uncertain situations and developed work groups to increase the chance of making effective decisions in the face of critical situations.

On a Leadership Imagination Retreat in June 2017 (Schwartz, 2017) at the University of Pennsylvania, a number of senior executives (CEOs, army generals, etc.) were asked, "What role does intuition play in your leadership practice?" This discussion centered on why leaders are not trained to be abductive thinkers. It was argued that most leaders are adept at inductive reasoning and deductive reasoning. It was suggested that leaders should be able to use abductive thinking to "leap to the right conclusion."

However, as we all know, intuition can have its downfalls. For example, a Dartmouth study showed that subjects' confidence had been disrupted by negative feedback, and they lost the relative accuracy advantage from relying on their intuition. Ma-Kellams and Lemer (2016) showed that in their study of executive-level professionals, those who relied on intuitive thinking also tended to exhibit lower empathic accuracy (i.e., the ability to infer the feelings of others, which is important for cultivating successful professional and personal relationships). Research by Kardes (2006) indicated that "managers should trust their intuition only when high-quality (frequent, prompt, and diagnostic) feedback is available and when inferential errors are consequential and therefore easy to detect." And, of course, in our age of big data and analytics, we continue to see a shift to more data-centric decision making and a lessening of the role of an executive's intuitive awareness (Ramrathan and Sibanda, 2017).

BECOMING A MORE INTUITIVE EXECUTIVE

Given the merits and potential pitfalls of applying intuitive awareness in executive decision making, it seems reasonable that a complementary set of both intuition and analytics should form the basis for making decisions. Maybe the secret formula for executives should be "Analytics + Intuition = Success."

In some of the seminal work by Sadler-Smith and Shefy (2004) on developing the *intuitive executive*, they suggest that there are ways to enhance an executive's intuition. Some of these techniques are as follows (Sadley-Smith and Shefy, 2004):

- Open up the closet: Perform some introspection on how well you trust your hunches when confronted by an important decision.
- Don't mix up your I's (intuition, insight, instinct): The *eureka* moment through incubation (the unconscious processing of information) typically results in intuition, which may then become validated as an insight.
- Elicit good feedback: An executive coach or mentor may help in this regard.
- Get a feel for your batting average: Benchmark how well your intuitive judgments have been in making past decisions.
- Use imagery: The more people are able to think visually, the more likely they are to use intuitive processes.
- Play devil's advocate: Challenge your assumptions; are you overconfident in your judgments?
- Capture and validate your intuitions: Perhaps keep a diary or document your intuition used in key decisions.

Other mind-expanding approaches have been used to increase intuitive awareness, including meditation/yoga, creativity-enhancing exercises, and being able to better listen to your body signals. According to the *Washington Post* (March 3, 2005), Jeffrey Abramson's Tower Companies in Bethesda, Maryland, "anyone who has been on the job at least three months can walk over to the Maharishi Peace Palace two blocks away for a four-day course on meditation. The classes are gratis, courtesy of Tower. Employees can even go on company time." In addition, many companies are offering free yoga classes at their company's fitness centers in order to increase relaxation and perhaps intuitive awareness.

INTUITION-BASED DECISION MAKING IN PRACTICE: INSIGHTS FROM EXECUTIVES IN THE FIELD

In the author's interview with Rear Admiral (Ret.) Norman Hayes on January 18, 2017, a discussion took place on how intuition and data/information play a role in strategic decision making. Admiral Hayes, who previously served as director of intelligence for the U.S. European Command, first shed light on his structured decision-making process. As an intelligence officer, Admiral Hayes says that he relies on information—that is, processed data—and essentially is interested in managing risk. He mentioned that if one is not experienced, the risk of not making the right decisions with the right outcomes will increase. He highlighted that intuition leads you in a direction partly based on predetermined experience. Essentially, he says that you "stay within boundaries," and you should always be assessing the risk in the decision that you will make. This is particularly true in a fluid environment, as one can imagine in a combat scenario.

Admiral Hayes quoted President Jimmy Carter: "I reserve the right to be inconsistent as I grow and change in knowledge and learning. A person who is stagnant and without growth is one who is unable to change or modify their opinion or accept others." This stresses the point that when leaders face new situations, when crises arise, they must continually ask questions of themselves to remain grounded in their experience and knowledge, logical as they proceed, rational in their analysis, and prepared to change and adapt rapidly in a fluid environment. Here, intuition plays a role as one needs to be prepared to reshape one's thoughts in real time.

Admiral Hayes related an episode from his early career in which he and his fellow junior officers coined the phrase "If you are flexible, you are too rigid. You have to be fluid." By way of further explanation, he noted that if you put a flexible hacksaw blade into a bottle or glass, the blade will bend and fold to fit inside the confined space, but greater than 90% of all the space in the bottle or glass is empty—useless volume. However, if you are fluid you will fill the complete volume; you will know, touch, and interact with your environment. You conform to the space available— releasing preconceived ideas, learning to be more receptive but with a heightened awareness of the environment. If you are intuitively aware, you are a fluid with variable viscosity, continually adapting to the morphing environment.

Admiral Hayes also mentioned the importance of falling back on your training and experience. He stated that even though intuition is often used in crisis situations, it is most helpful to have prepared in advance. He referred to the creation of a crisis action plan while serving at the U.S. European Command. The idea is to prepare for potential crisis situations. If you build and then practice executing a crisis action plan, if a crisis occurs, you will be able to act more decisively in the early stages of the crisis, making directive decisions and reducing intuitive decisions. By doing so, you create space and time to think and react to evolving situations in the crisis. The thought is to become an adaptive thinker using practiced processes in responding to new situations.

In closing, Admiral Hayes echoed that a blend of intuitive thought (the truncation of individual decision processes), rapid outcome analysis, and risk management are essential for constrained time lines in a decision-making environment. The key to making effective decisions is to be adaptive, fluid, and idea receptive in a rapidly morphing environment in the pursuit of shaping outcomes.

In another author interview with Dan Ranta, the executive knowledge sharing and library/research leader for GE-Wide (and past Knowledge Leader for a major oil and gas company), on January 19, 2018, Dan mentioned that he relies on his intuition and past experience in a number of ways. At GE, Dan's team has launched online communities of practice in various business segments, with about 89,000 total community memberships and growing in just the 2 years since Dan's team started the program at GE. According to his strategic road map, there are currently 139 communities at GE, and they are heading toward about 225 communities by the end of 2018.

Dan's intuition and experiential learning has taught him that instilling *trust* is the key factor in order to develop, nurture, and gain work-related value from these online communities. He highlighted the product lifecycle management (PLM) area, which deals with the ability to handle logistics, parts inventories, and the like in the various GE businesses (e.g., aviation, oil and gas, etc.). Dan knew that he couldn't get these communities off the ground without building trust within the communities. He instilled trust by listening to the needs of potential community leaders and members, being empathetic, and sending a message of sincerity. His intuitive feel has resulted in key performance indicator (KPI) enhancements throughout the various communities. As examples, they now have about 1100 members in the Aviation community, 2150 members in the oil and gas community, and even nearly 900 members in the *knowledge-sharing* community (where, through crowdsourcing, members share ideas about how best to do knowledge sharing at GE). According to Dan, 75% of our best ideas come from our customers and most of those by way of the knowledge-sharing community.

In asking Dan about where his intuition may have failed him, Dan discussed an example where he set up an online community, during his ConocoPhillips days, in western Canada (around the Calgary area). There were about 200 people in this online community, but unfortunately it failed because there wasn't a big enough global scope, and those community members already knew each other fairly well, so it was easier just to use the standard email and telephone. The community also lacked effective management support. Dan initially thought this would be a successful community of practice, even though his colleagues at the time forewarned him. Dan said, in hindsight, that it's good to have some swagger but not be overconfident.

Dan runs about 4 miles every day on the treadmill to help his thought process and perhaps his intuitive awareness. While on the treadmill, he thinks about the "edges" in terms of thinking "big, but in bite-size pieces where strategy can move efficiently to action." He is constantly looking for learning opportunities across the enterprise, and exchanges ideas for improving his knowledge-sharing strategy with colleagues inside and outside GE. He wrote 60 blogs last year and believes in *relentless messaging* and *relentless coaching* to focus on KPIs and drive excellence for this digital industrial company.

SOME CLOSING THOUGHTS

Intuition and analytics form a winning combination, but in this data-driven society, experiential learning and "gut feel" are often overlooked. In General Colin Powell's (former chairman, Joint Chiefs of Staff) "Leadership Primer," he states:

Part 1: Use the formula $P = 40$ to 70, in which P stands for the probability of success and the numbers indicate the percentage of information acquired.

Part 2: Once the information is in the 40 to 70 range, go with your gut.

Don't take action if you have only enough information to give you less than a 40 percent chance of being right, but don't wait until you have enough facts to be 100 percent sure, because by then it is almost always too late. Today, excessive delays in the name of information-gathering breeds "analysis paralysis." Procrastination in the name of reducing risk actually increases risk.

This chapter highlights the pros and cons of using intuition and presents some of the evidence-based research and examples from executives in applying their intuitive awareness. The rest of this chapter, as shown in Appendix A, presents summaries of much of the leading research, over the recent years, that has been published in this intuition-based executive decision–making realm. Hopefully, analytics professionals and senior executives will make intuition a part of their toolkit in making key decisions.

APPENDIX A: ANNOTATED INTUITION STUDIES, WITH SPECIAL EMPHASIS ON EXECUTIVE DECISION MAKING

C. Ma-Kellams and J. Lerner (2016), "Trust Your Gut or Think Carefully? Examining Whether an Intuitive, versus a Systematic, Mode of Thought Produces Greater Empathic Accuracy," Faculty Research Working Paper Series, Harvard Kennedy School, RWP 16-017, April.

Using executive-level professionals as participants, the studies showed that people who tend to rely on intuitive thinking also tend to exhibit lower empathic accuracy. Contrary to lay beliefs, empathic accuracy (e.g., the ability to accurately infer the feelings of others) arises more from systematic thought than from gut intuition. Used 72 international and U.S.-born executives.

J. Pretz, J. Brookings, L. Carlson, T. Humbert, M. Roy, M. Jones, and D. Memmert (2014), "Development and Validation of a New Measure of Intuition: The Types of Intuition Scale," *Journal of Behavioral Decision Making*, Vol. 27.

Developed the Types of Intuition Scale based on three distinct types of intuition:

Holistic intuitions: Judgments based on a qualitatively non-analytical process (Holistic–Big Picture, Holistic–Abstract)
Inferential intuitions: Judgments based on automated inferences, decision-making processes that were once analytical but have become intuitive with practice
Affective intuitions: Judgments based primarily on emotional reactions to decision situations

N. Kandasamy, S. Garfinkel, L. Page, B. Hardy, H. Critchley, M. Gurnell, and J. Coates (2016), "Interoceptive Ability Predicts Survival on a London Trading Floor," *Scientific Reports*, September 19.

Financial traders are better at reading their "gut feelings" than the general population; and the better they are at this ability, the more successful they are as traders. $N = 18$ male traders. The trader's heartbeat counting score predicted the number of years s/he had survived as a trader.

P. Hassani, A. Abdi, and R. Jalali (2016), "State of Science, 'Intuition in Nursing Practice': A Systematic Review Study," *Journal of Clinical and Diagnostic Research.*

A systematic review study (144 articles with "intuition and nursing") revealed that research about intuition in nursing was mostly conducted with qualitative research methodologies, and there is a lack of quantitative and trial studies.

G. Klein (2015), "A Naturalistic Decision Making Perspective on Studying Intuitive Decision Making," *Journal of Applied Research in Memory and Cognition*, Vol. 4.

Intuitions can be strengthened by providing a broader experience base that lets people build better tacit knowledge, such as perceptual skills and richer mental models, as a means of achieving better decisions.

Y. Wang, S. Highhouse, C. Lake, N. Petersen, and T. Rada (2015), "Meta-analytic Investigations of the Relation between Intuition and Analysis," *Journal of Behavioral Decision Making.*

Their findings support the view that intuition and analysis are independent constructs rather than opposite ends of a bipolar continuum. In addition, the findings suggest measures of analysis or rationality are not interchangeable.

V. Thoma, E. White, A. Panigrahi, V. Strowger, and I. Anderson (2015), "Good Thinking or Gut Feeling? Cognitive Reflection and Intuition in Traders, Bankers, and Financial Non-Experts," *PLOS One*, April.

Traders showed no elevated preference to use "intuition" in their decision making compared with other groups. Results indicate that compared to non-expert participants, financial traders have a higher self-rated tendency for reflective thinking and a greater propensity to inhibit the use of mental shortcuts (heuristics) in decision making.

P. Hanlon (2011), "The Role of Intuition in Strategic Decision Making: How Managers Rationalize Intuition," Dublin Institute of Technology, 14th Annual Conference of the Irish Academy of Management, September.

$N = 12$ senior managers in Ireland were interviewed. Research identified that intuition is used by managers in strategic decision making. A conceptual Deconstructing Intuition Model was developed as follows: *manifesting intuition* (feelings, location) → *aspects of intuition* (anchor, early warning system, dual role, use in vision) → *externalizing and rationalizing intuition* (testing intuitive hypotheses, making a rational case, deliberately ignoring).

M. Usher, Z. Russo, M. Weyers, R. Brauner, and D. Zakay (2011), "The Impact of the Mode of Thought in Complex Decisions: Intuitive Decisions Are Better," *Frontiers in Psychology*, March 15.

Intuitive or affective manipulations improve decision quality compared with analytic/deliberation manipulations (without distraction). $N = 36$ executives.

J. Paliszkiewicz, J. Gołuchowski, and A. Koohang (2015), "Leadership, Trust, and Knowledge Management in Relation to Organizational Performance: Developing an Instrument," *Online Journal of Applied Knowledge Management*, Vol. 3, No. 2.

Develops a trust management instrument to measure the effect of trust management on knowledge management and organizational performance. The trust management construct looks at 10 characteristics: ability/competence, benevolence, communication, congruency, consistency, dependability, integrity, openness, reliability, and transparency.

J. Moxley, K. Ericsson, N. Charness, and R. Krampe (2012), "The Role of Intuition and Deliberative Thinking in Experts' Superior Tactical Decision Making," *Cognition*, Vol. 124.

Both experts and less skilled individuals benefit significantly from extra deliberation, regardless of whether the problem is easy or difficult.

A. Koohang, J. Paliszkiewicz, and J. Gołuchowski (2017), "The Impact of Leadership on Trust, Knowledge Management, and Organizational Performance: A Research Model," *Industrial Management and Data Systems Journal,* **Vol. 117, No. 3.**

The study's findings showed positive and significant linear connection among leadership, trust, knowledge management, and organizational performance.

J. Woiceshyn (2009), "Lessons from 'Good Minds': How CEOs Use Intuition, Analysis and Guiding Principles to Make Strategic Decisions," *Long Range Planning Journal,* **Vol. 42, Elsevier.**

The most effective CEOs used integration by essentials and spiraling. They also shared three thinking-related traits: focus, motivation, and self-awareness. $N = 19$ oil company CEOs.

C. Akinci and E. Sadler-Smith (2012), "Intuition in Management Research: A Historical Review," *International Journal of Management Reviews,* **Vol. 14.**

Intuitive expertise is a growing research area that conjoins two major traditions in intuition research: namely, *naturalistic decision making* (NDM) and *heuristics and biases.* Also, the neuroscience of intuition (e.g., fMRI, etc.) will be a growing area to examine the neural bases of intuitive judgment and its associated processes.

K. Martin, M. Kebbell, L. Porter, and M. Townsley (2011), "The Paradox of Intuitive Analysis and the Implications for Professionalism," *Journal of the Australian Institute of Professional Intelligence Officers,* **Vol. 19, No. 1.**

Analysts are still not fully conscious of their role as decision makers.

B. Violino (2014), "Do Executives Trust Data or Intuition in Decision Making?," *Information Management,* **June 9.**

A study by the Economist Intelligence Unit (commissioned by Applied Predictive Technologies) showed that nearly three-quarters of the executives surveyed said they trust their own intuition when it comes to decision making, and 68% believed they would be trusted to make a decision that was not supported by data.

N. Cameron (2014), "Report: Customer Data Analysis Is Up, but Execs Still Rely on Intuition for Strategic Decisions," CMO, September 15 (www.cmo.com.au).

The EIU and PwC report "Gut & Gigabytes: Capitalizing on the Art and Science in Decision Making" claims four out of five Australian executives are using data and analysis to optimize customer value for their organization, even though most still rely on their intuition or peer advice to make strategic business decisions. Of the Australian respondents, 80% claimed they use data and analysis around customer value; in contrast, just 11% of U.S. respondents said they do the same. Of global C-suite executives, 52% admitted to previously discounting data they didn't understand and 31% said the timeliness of data was an issue. $N = 1135$ executives (54% of which were C-level or board members).

J. Hedge and K. Aspinwall (2009), "Can Intuitive Decision Making Improve Homeland Security?," Institute for Homeland Security Solutions Research Brief, November.

There may be substantial promise in the application of intuitive decision-making strategies for homeland security, especially when dealing with extreme time pressures.

S. Swami (2013), "Executive Functions and Decision Making: A Managerial Review," *IIMB Management Review*, Vol. 25.

Many decisions are made unconsciously in our mind, such as situations with higher time pressures, higher stakes, or increased ambiguities. In these situations, experts may well use intuitive decision making rather than structured approaches.

M. Andrzejewska, D. Berkay, S. Dreesmann, J. Haslbeck, D. Mechelmans, and S. Furlan (2013), "(In)accurate Intuition: Fast Reasoning in Decision Making," *Journal of European Psychology Students*.

It is hypothesized that people with high expertise levels will perform equally well both under and without time pressure, and people higher on cognitive abilities will perform equally well under time pressure and without time pressure, unlike those with lower cognitive abilities. Research was proposed.

KPMG (2016), "The Trust Gap: Building Trust in Analytics," Report, October 31.

Forrester Consulting was commissioned by KPMG to survey 2165 respondents from 10 countries. Most business leaders today believe in the value of data and analytics (D&A) but say they lack confidence in their ability to measure the effectiveness and impact of D&A and mistrust the analytics used to help drive decision making. Seventy percent of leaders believe using data and analytics can expose their organizations to reputational risk. Only 11% have trust in using/deploying analytics.

R. Davis-Floyd and P. Arvidson (ed.) (1997), *Intuition: The Inside Story; Interdisciplinary Perspectives*, Princeton University, NJ.

The U.S. Marine Corps' Command and Control Vision stresses intuition as central to victory. Bell Atlantic Corporation listed intuition as an important quality on job descriptions, and the role of intuition in the successful R&D process has been highlighted.

T. King (2016), "KPMG Study: Just One-Third of CEOs Trust Data Analytics," *Solutions Review*, July 20.

Only one-third of CEOs have a high level of trust in the accuracy of their data analytics. $N = 400$ chief executives. Seventy-seven percent polled said that they had concerns about internal data quality.

M. Mannor, A. Wowak, V. Bartkus, and L. Gomez-Mejia (2016), "How Anxiety Affects CEO Decision Making," *Harvard Business Review*, July 19.

$N = 84$ CEOs interviewed. More-anxious leaders took fewer strategic risks than their less anxious peers in order to avoid potential losses.

S. Hammons (2015), "Navy Research Project on Intuition," www. cultureready.org, April 6.

The Office of Naval Research (ONR) funded a 4-year project to identify, understand, and use intuitive decision making (implicit learning). This was launched in 2014 (4-year, $3.85 million). A seasoned warfighter develops a gut instinct through experience.

Z. Tormala, T. Riesterer, E. Peterson, and C. Smith (n.d.), "Losses and Gains: Does Loss Aversion Influence Executive Decision Making?," Corporate Visions, Inc.

$N = 113$ executives in an online experiment. Results show that loss aversion influences business and personal decisions in a sample of executives. Executive decision makers are not as rational as might be imagined by marketers and salespeople.

M. Sinclair, N. Ashkanasy, and P. Chattopadhyay (2010), "Affective Antecedents of Intuitive Decision Making," *Journal of Management & Organization*, Vol. 16, No. 3, July.

Emotional awareness has a positive effect on the use of intuition, which appears to be stronger for women. Surprisingly, positive and negative moods seem to influence intuition according to their intensity rather than positive/negative distinction.

R. Hogarth (2010), "Intuition: A Challenge for Psychological Research on Decision Making," *Psychological Inquiry*, Vol. 21.

Intuition is shaped by learning. As tasks become more analytically complex, the advantage shifts to intuition (but this is subject to the provision that the decision maker's intuitions have been honed in kind environments). Perhaps there may be a rationale for delaying decisions by "sleeping on them," thereby allowing unconscious thought (intuition?) to improve them.

E. Sadler-Smith (2016), "'What Happens When You Intuit?': Understanding Human Resource Practitioners' Subjective Experience of Intuition Through a Novel Linguistic Method," *Human Relations*, Vol. 69, No. 5, May.

$N = 124$ HR practitioners. This research uncovers two aspects of intuitive affect: bodily awareness and cognitive awareness.

K. Leavitt, L. Zhu, and K. Aquino (2016), "Good Without Knowing It: Subtle Contextual Cues Can Activate Moral Identity and Reshape Moral Intuition," *Journal of Business Ethics*, Vol. 137.

Subtle contextual cues can lead individuals to render more ethical judgments by automatically restructuring moral intuition below the level of consciousness.

K. Malewska (2015), "Intuition in Decision Making: Theoretical and Empirical Aspects," *Journal of the Academy of Business and Retail Management*, Vol. 3, No. 3, November.

$N = 48$ managers. All respondents said they treat intuition as an important part of any decision making process. The quasi-intuitive approach prevailed among top-level managers.

V. Cavojova and R. Hanak (2014), "How Much Information Do You Need? Interaction of Intuitive Processing with Expertise," *Studia Psychologica*, Vol. 56.

Results indicate that situational manipulations, such as inducing time stress or giving instructions to think intuitively, affect information searches more than the preferred cognitive style and that it's necessary to examine intuition in context-specific tasks, as the experience plays a crucial role in searching information when making decisions.

M. Wright (2013), "Homicide Detectives' Intuition," *Journal of Investigative Psychology and Offender Profiling*, Vol. 10.

Forty detectives made 594 inferences, of which 67% ($N = 398$) were accurate. A homicide detective's intuition is a cognitive skill that stems from the experience of investigating homicide. The ability to draw inferences and make decisions from crime scene information is an important skill for detectives to develop.

D. Officer (2005), "The Unexplored Relationship between Intuition and Innovation," *The Innovation Journal: The Public Sector Innovation Journal*, Vol. 10, No. 3.

Intuition in tune with innovation is often a catalyst instrumental in prompting the very best inspiration. Intuition can be "fine-tuned."

G. Lufityanto, C. Donkin, and J. Pearson (2016), "Measuring Intuition," *Psychological Science*, Vol. 27, No. 5, May.

The study's behavioral and physiological data, along with evidence-accumulator models, show that non-conscious emotional information can boost accuracy and confidence in a concurrent emotion-free decision task, while also speeding up response times. A model that simultaneously accumulates evidence from both physiological skin conductance and conscious decisional information provides an accurate description of the data. The findings support the notion that non-conscious emotions can bias concurrent non-emotional behavior—a process of intuition.

D. Clarke and A. Hunt (2016), "Failure of Intuition: When Choosing Whether to Invest in a Single Goal or Split Resources between Two Goals," *Psychological Science*, Vol. 27, No. 1, January.

The majority of the participants consistently failed to modify their strategy in response to changes in task difficulty. This failure may have been related to uncertainty about their own ability.

E. Mikuskova, R. Hanak, and V. Cavojova (2015), "Appropriateness of Two Inventories Measuring Intuition (The PID and the REI) for Slovak Population," *Studia Psychologica*, Vol. 57.

Results showed that both the Preference for Intuition/Deliberation (PID) and the Rational-Experiential Inventory (REI) have good internal consistency. They also examined underlying factors behind scales for measuring intuition—decision making based on affect and holistic processing, decision making based on creativity and cognitions, and planned, deliberative decision making. The analyses conclude that the scales used differentiate appropriately between intuitive and deliberative cognitive styles, and they probably measure different facets of the same underlying construct. $N = 428$ working adults and students.

D. Petrovic, M. Illic, S. Stojanovic, and V. Stoickov (2014), "Acute Myocardial Infarction with an Initially Non-diagnostic Electrocardiogram: Clinical Intuition is Crucial for Decision Making," *Scientific Journal of the Faculty of Medicine in Nis*, Vol. 31, No. 3.

The case highlights the importance of using clinical intuition for decision making.

C. Schmidt (2014), "Questioning Intuition through Reflective Engagement", *Journal of Moral Education*, Vol. 43, No. 4.

The findings suggest that reflective engagement enables the study's participants to become more aware of and therefore *access* and *govern* intuitions so that they can be more equally integrated during a moral decision-making process.

K. Rauf (2014), "Use of Intuition in Decision Making among Managers in Banking and Industrial Sectors of Karachi," *Pakistan Journal of Psychological Research*, Vol. 29, No. 1.

$N = 306$ managers. Respondents reported the use of intuition by people who work under intense pressure (25.6% and 29.8%, respectively, in the industrial and banking sectors). It's mentioned that Mintzberg (McGill) indicates that entrepreneurs generally rely on intuitive decision making or they rest their decisions on realities that confirm their intuition only. Fifty-six percent of respondents prefer to rely on experience from the industrial sector and 46.4% from banking sector.

C. Remmers and J. Michalak (2016), "Losing Gut Feeling? Intuition in Depression," *Frontiers in Psychology*, August.

In contrast to healthy individuals who take most daily life decisions intuitively (Kahneman), depressed individuals seem to have difficulty coming to fast and adaptive decisions.

J. Okoli, G. Weller, and J. Watt (2016), "Information Processing and Intuitive Decision Making on the Fireground: Towards a Model of Expert Intuition," *Cognition, Technology & Work*, Vol. 18, No. 1, February.

$N = 16$ experienced fireground commanders were interviewed. An information-filtering and intuitive decision-making model is proposed. The model attempts to conceptualize how experienced firefighters scan through multiple information sources from which they are then able to select the most relevant cues that eventually aid the development of workable action plans.

W. van Riel, J. Langevel, P. Herder, and F. Clemens (2014), "Intuition and Information in Decision Making for Sewer Asset Management," *Urban Water Journal*, Vol. 11, No. 6.

Given the complex context of sewer asset management and limited data, intuitive decision making is favored but is, however, "not skilled" (because the two conditions for intuition to be skilled—sufficient regularity and learning opportunity—are not met).

A. Nita and I. Solomon (2015), "The Role of Intuition and Decision Making in Public Administration," *Juridical Current*, Vol. 61, June.

In public administration, intuitive ability plays an important role when public officials are facing new situations. It's important to be an *intuitive leader*.

E. Margolis and S. Laurence (2003), "Should We Trust Our Intuitions? Deflationary Accounts of the Analytic Data," *Proceedings of the Aristotelian Society*, Wiley.

Intuitions of analyticity are intuitions that seem to reflect truths of meaning. This article argues that opponents of analyticity have some unexpected resources for explaining these intuitions and that, accordingly, the argument from intuition fails.

J. Andow (2015), "How 'Intuition' Exploded," *Metaphilosophy*, Vol. 46, No. 2, April.

The number of philosophy papers published that make explicit appeal to intuition has markedly increased over the past 10 years.

S. Biddulph (2015), "Guiding Principles in Rational and Intuitive Strategic Decision Making at a Chemicals Business," Gordon Institute of Business Science Research Report (in partial fulfillment of MBA degree), University of Pretoria, South Africa, November 9.

As experience and knowledge escalate over time, the rational decision-making approach becomes more intuitive, allowing the decision-making process to become quicker and the possibility of lost opportunities to lessen.

O. Hyppanen (2013), "Decision Makers' Use of Intuition at the Front End of Innovation," doctoral dissertation, Department of Industrial Engineering, Aalto University, Finland.

Findings in management decision-making research suggest that decision makers often use intuition in uncertain situations. $N = 19$ experienced decision makers (interviews). $N = 86$ experienced and inexperienced decision makers surveyed. Findings suggest that intuition plays a major role for experienced decision makers when making innovation front-end decisions. The results also stress that the role intuition plays should receive greater acknowledgment in innovation research.

R. Quammen (2015), "Intuition and Its Impact on Information Systems," DBA dissertation, Kennesaw State University, GA, April 21.

This research introduces IS intuition as a mediating variable impacting IS (Information System) success. 62% of healthcare executives, managers, and IS consultants believe intuition impacts IS success and EHR (Electronic Health Record) adoption.

J. Mikels, S. Maglio, A. Reed, and J. Kaplowitz (2011), "Should I Go with My Gut: Investigating the Benefits of Emotion-Focused Decision Making," *Emotion*, American Psychological Association, May.

Faculty from four universities, including Stanford, conducted four studies involving over 230 students. Participants were asked to make decisions based on a feeling-focused strategy, a reason-focused strategy, and no specific strategy. The results from the studies indicated that focusing on feelings versus details led to superior objective and subjective decision quality for complex decisions.

G. Hofstede, "Cultural Dimensions," https://geerthofstede.com

Based on Hofstede's cultural dimensions, the following countries scored as follows:

1. Power distance: Italy (50), Poland (68), Canada (39), United States (40)
2. Individualism: Italy (76), Poland (60), Canada (80), United States (91)
3. Masculinity: Italy (70), Poland (64), Canada (52), United States (62)
4. Uncertainty avoidance: Italy (75), Poland (93), Canada (48), United States (46)
5. Long-term orientation: Italy (61), Poland (38), Canada (36), United States (26)
6. Indulgence: Italy (30), Poland (29), Canada (68), United States (68)

A. Norenzayan, E. Smith, B. Kim, and R. Nisbett (2002), "Cultural Preferences for Formal versus Intuitive Reasoning," *Cognitive Science*, Vol. 26, No. 5, September–October.

European Americans, more than Chinese and Koreans, set aside intuition in favor of formal reasoning. Conversely, Chinese and Koreans relied on intuitive strategies more than European Americans. Asian Americans' reasoning was either identical to that of European Americans or intermediate.

E. Buchtel and A. Norenzayan (2008) "Which Should You Use, Intuition or Logic? Cultural Differences in Injunctive Norms about Reasoning," *Asian Journal of Social Psychology*, Vol. 11.

European Canadians and East Asian Canadians read scenarios of intuitive versus rule-following business decisions. Relative to Western participants, East Asians rated intuitive reasoning as more important and reasonable than analytic reasoning.

GLOBE CEO Study (2014)

Data was collected from over 1000 CEOs and over 5000 senior executives in corporations in a variety of industries in 24 countries. The findings reinforce the importance of CEOs to organizational outcomes, the considerable influence of culture on societal leadership expectations, and the importance of matching CEO behaviors to the leadership expectations within each society. High-performing CEOs were visionary, team integrators, performance-oriented, administratively competent, decisive, and inspirational compared with underperforming CEOs.

J. Hill, A. Puurula, A. Sitko-Lutek, and A. Rakowska (2000), "Cognitive Style and Socialisation: An Exploration of Learned Sources of Style in Finland, Poland, and the UK," *Educational Psychology*, Vol. 20, No. 3.

$N = 200$ managers in Finland, Poland, and the United Kingdom. Relative preference for intuition is positively related to the level of the respondent's work position. Non-managers are more analytical than managers. Female managers are more intuitive than males in both the managerial and "others" groups.

C. Schreier, A. Schubert, J. Weber, and F. Farrar (2014), "An Investigation of the Character Traits of Decision-Makers Open to Intuition as a Tool," *GSTF Journal on Business Review*, Vol. 3, No. 4, November.

$N = 159$. Results show that leaders open to intuition are also open to new experiences.

J. Hayes, C. Allinson, and S. Armstrong (2004), "Intuition, Women Managers and Gendered Stereotypes," *Personnel Review*, Vol. 33, No. 4.

Results showed there is no difference between female and male managers in terms of intuitive orientation.

K. Matzler, F. Bailom, and T. Mooradian (2007), "Intuitive Decision Making," *MIT Sloan Management Review*, Vol. 49, No. 1, Fall.

Executives can hone their intuition by experience, networks, emotional intelligence, tolerance, curiosity, and limits.

I. Erenda, M. Mesko, and B. Bukovec (2014), "Intuitive Decision Making and Leadership Competencies of Managers in Slovenian Automotive Industry," *Journal of Universal Excellence*, **Vol. 3, No. 2, June.**

Intuitive decision making by top and middle managers in the auto industry can be characterized by their ability to recognize emotions.

Worthy Intuition-Related Books of Note:

- D. Kahneman (2011), *Thinking Fast and Slow*, Farrar, Straus, and Giroux.
- M. Gladwell (2005), *Blink*, Penguin.
- J. Liebowitz (ed.) (2014), *Bursting the Big Data Bubble: The Case for Intuition-Based Decision Making*, CRC Press, Boca Raton, FL.
- M. Lewis (2017), *The Undoing Project*, W.W. Norton.
- A. Koohang and J. Paliszkiewicz (2016), *Social Media and Trust*, Informing Science Press.
- M. Sinclair (ed.) (2014), *Handbook of Intuition Research*, Edward Elgar.
- J. Liebowitz, J. Paliszkiewicz, and J. Gołuchowski (eds.) (2018), *Intuition, Trust, and Analytics*, CRC Press, Boca Raton, FL.

REFERENCES

Fox, J., W. Wagedorn, and S. Sivo (2016), "Clinical decision-making and intuition: A task analysis of 44 experienced counsellors," *Counselling and Psychotherapy Research*, Vol. 16, No. 4, December.

Kardes, F. (2006), "When should consumers and managers trust their intuition?," *Journal of Consumer Psychology, Lawrence Erlbaum Associates*, Vol. 16, No. 1.

Liebowitz, J., J. Paliszkiewicz, and J. Gołuchowski (2018), *Intuition, Trust, and Analytics*, CRC Press, Boca Raton, FL.

Ma-Kellams, C. and J. Lemer (2016), "Trust your gut or think carefully? Examining whether intuitive, versus a systematic, mode of thought produces greater empathic accuracy," *Journal of Personality and Social Psychology*, Vol. 111, No. 5, November.

Marinos, G., and N. Rosni (2017), *"The Role of Intuition in Executive Strategic Decision Making,"* master's thesis, School of Economics and Management, Lund University, May.

Moore, J. (2017), "Data visualization in support of executive decision making," *Interdisciplinary Journal of Information, Knowledge, and Management*, Vol. 12, Informing Science Institute.

Ramrathan, D., and M. Sibanda (2017), "The impact of information technology advancement on intuition in organizations: A phenomenological approach," *Journal of Developing Areas*, Vol. 51, No. 1, Winter.

Rauf, K. (2014), "Use of intuition in decision making among managers in banking and industrial sectors of Karachi," *Pakistan Journal of Psychological Research*, Vol. 29, No. 1, Summer.

Robert, R., D. Tilley, and S. Petersen (2014), "A power in clinical nursing practice: Concept analysis on nursing intuition," *MEDSURG Nursing*, Vol. 23, No. 5, September–October.

Rovithis, M., A. Stavropoulou, N. Katsigaraki, M. Sotiropoulos, D. Sfigkaki, M. Linardakis, and N. Rikos (2015), "Evaluation of intuition levels in nursing staff," *Health Science Journal*, Vol. 9, No. 3–4.

Sadler-Smith, E., and E. Shefy (2004), "The intuitive executive: understanding and applying 'gut feel' in decision-making," *Academy of Management Executive*, Vol. 18, No. 4.

Schwartz, A. (2017), "Report on the Leadership Imagination Retreat," Imagination Institute, University of Pennsylvania, Philadelphia, June 23.

2

How Gender, Experience, Role, Industry, and Country Affect Intuition

Written by: Yolande Chan, Tracy A. Jenkin, and Dylan Spicker

CONTENTS

INTRODUCTION AND FINDINGS

Recent work has suggested that the depiction of intuition as a single, unitary process is misguided. In 2014, the Types of Intuition Scale (TIntS) was proposed, which characterized intuition as four separate and independent constructs that could be measured using a survey instrument (Pretz et al., 2014). Instead of speaking directly of intuition, the paper (building on previous research; Pretz and Totz, 2007) discussed *Affective, Inferential, Holistic Abstract,* and *Holistic Big Picture* intuition. Affective intuition refers to judgments that are made through predominantly emotional mechanisms. Inferential Intuition refers to judgments made through embedded analytical processes that have become automatic through experience. Holistic Intuition, broadly speaking, concerns judgments that are made through the synthesis of diverse cues and are fundamentally non-analytical. Holistic Abstract Intuition bases judgments on a theoretical worldview, whereas Holistic Big Picture Intuition leverages a full-systems approach to the synthesis of diverse information.

The existence of these diverse processes, which are all categorized as *intuitive thought,* leads to questions regarding the use of intuition by individuals in the decision-making process. When we talk of an executive using intuition to "know" the best strategic path for the company versus a manager who leverages his or her intuition to guide a team to success, are we talking about the same process? Are managers in various industries or positions aided by a particular intuitive approach or hindered by another? Are there times when analysis is less well-suited to address a concern when compared with intuition? Are there areas where analysis and intuition are, in fact, two sides of the same coin?

The research that follows attempts to build on the work done by Pretz et al. (2014), seeking to tease apart differences that exist in the underlying intuitive process across workers in various environments. It contrasts employees with varying positions, from varying industries, with a diverse range of experience, and seeks to determine the interplay of analysis and intuition in the decision-making processes of these individuals.

The study findings confirm the existence of the constructs proposed by Pretz et al. (2014), though modification to the instrument was required. Additionally, the study finds distinct patterns in the use of the varying intuitive processes. In particular, individuals who rely heavily on Holistic Abstract Intuition (the process through which judgments are made based on a theoretical framework) also reliably trust in their Affective Intuition (the process through which emotional responses drive judgments). Conversely, individuals who rely on Holistic Big Picture Intuition (the process wherein judgments are made through a full-systems approach) also favor Inferential Intuition (previously analytical processes that have become second nature). These distinct "styles" of intuition remain when controlling for covariates and appear to be a consistent trend in the sample. We refer to the first group, who trust in emotional and theoretical processes, as *Feeling Theorists*, and the second group, relying on big picture and experience, as *Big Picture Modelers*.

The study also finds strong evidence for trends in the use of the preceding styles. The intuitive processes are dynamic, changing with experience. Less experienced workers are more likely to be Feeling Theorists when compared with their more experienced colleagues, who are more likely to be Big Picture Modelers. This trend remains when controlling for the positions that these workers hold. This dynamic trend is far and away the most prominent result, though minor static trait differences remain, even when controlling for experience. Managers are prone to rely on Affective Intuition when compared with C-suite executives; some industries have unique profiles, differentiating them from other industries; small differences also exist across countries. To further enhance the results that can be drawn, the study then weaves in the use of analytical decision-making processes, overlaying these results. Through this, a holistic view of judgment mechanisms is presented, providing fertile ground for further questions.

THE STUDY: DETAILS

A web-based survey was employed, collecting 172 responses from Canada, the United States, Poland, and Italy. Respondents were diverse in the industries they represented as well as the positions they held (C-suite, directors, managers, and staff). To accommodate the diverse range of industries represented, the responses were standardized to fit one of seven possible industries (Table 2.1).

Respondents were asked to rate their level of agreement (on a five-point Likert scale: 5 represents strongly agree and 1 represents strongly disagree) regarding statements about how they make decisions. A significant portion of the questions were taken, verbatim, from the work done by Pretz, using the proposed TIntS instrument (Pretz et al., 2014).

Following the data collection, a confirmatory factor analysis (CFA) was used to validate that the survey results were consistent with existing theory. There existed some differences between the existing instrument

TABLE 2.1

Description of the Industry Groupings, Used for Analysis

Industry Title	Description
Not for profit (NFP)	Refers to any individuals working for NFPs/charities, in addition to government workers, and those in education.
Marketing	Refers to any workers in marketing, advertising, communications, or a related field.
Engineering and Manufacturing	Refers to workers in fields related to manufacturing, engineering, utilities, transportation, etc.
Professional Services	Refers to those working explicitly in professional services (i.e., consulting) or those working in finance and other related industries.
Information Technology (IT)	Refers explicitly to those working in the technology field.
Pharma	Refers to workers in pharmacy or in tangential fields, including biotech and research and development.
Agriculture	Refers to individuals specifically working in farming and agriculture. (*Note*: due to $N = 1$, this industry group was omitted from most analyses).

and the mappings that were found to be significant. Specifically, a trimmed model of 15 items (reduced from the initial 23-item scale) generated results consistent with strong model fit (CFI 0.872; RMSEA 0.063; SRMSR 0.079). The instrument that was used to assess intuition, and how it differs from previous work, is shown in Table 2.2.

The original instrument proposed by Pretz et al. (2014) did not include items for Analysis. To develop a proxy metric, it was theorized that the reverse-worded questions in the instrument could form a composite measure that would reflect an individual's propensity toward Analysis. Running Velicer's (1976) MAP test on the reverse-worded questions suggests that they all map onto a single factor. (The MAP test was run using the R package *paramap* version 1.3; O'Connor, 2017.) This suggests that the reverse-worded items correspond to a single aggregate factor. As such, the eight reverse-worded questions were taken as a proxy for an employee's tendency toward analytical judgment (Table 2.2).

A simple average was computed to determine employees' preferences for the various constructs. Taking these preferences, comparisons were then made across the various demographic variables that had been collected in the survey; a breakdown of the sample along these factors can be seen in Figure 2.1.

TABLE 2.2

Complete Survey Proposed by Pretz et al. (2014), with Corresponding Mappings for Tint23 and Findings from the CFA

Survey Question	Tint23[a] (Pretz et al., 2014)	CFA Result
1. When tackling a new project, I concentrate on big ideas rather than the details.	HB	Omitted
2. I trust my intuitions, especially in familiar situations.	I	I
3. I prefer to use my emotional hunches to deal with a problem rather than thinking about it.	A	A
4. Familiar problems can often be solved intuitively.	I	I
5. It is better to break a problem into parts than to focus on the big picture.	Omitted	Analysis
6. There is a logical justification for most of my intuitive judgments.	I	I
7. I rarely allow my emotional reactions to override logic.	(R) A	Analysis
8. My approach to problem solving relies heavily on my past experience.	Omitted	Omitted
9. I tend to use my heart as a guide for my actions.	A	A
10. My intuitions come to me very quickly.	I	I
11. I would rather think in terms of theories than facts.	HA	HA
12. My intuitions are based on my experience.	I	Omitted
13. I often make decisions based on my gut feelings, even when the decision is contrary to objective information.	A	A
14. When working on a complex problem or decision, I tend to focus on the details and lose sight of the big picture.	(R) HB	Analysis
15. When making decisions, I value my feelings and hunches just as much as I value facts.	Omitted	Omitted
16. I believe in trusting my hunches.	A	A
17. When I have experience or knowledge about a problem, I trust my intuitions.	Omitted	Omitted
18. I prefer concrete facts over abstract theories.	(R) HA	(R) HA[b] Analysis
19. When making a quick decision in my area of expertise, I can justify the decision logically.	I	I
20. I generally don't depend on my feelings to help me make decisions.	(R) A	Analysis

(Continued)

TABLE 2.2 (CONTINUED)

Complete Survey Proposed by Pretz et al. (2014), with Corresponding Mappings for Tint23 and Findings from the CFA

Survey Question	Tint23 (Pretz et al., 2014)	CFA Result
21. I've had enough experience to know what I need to do most of the time without trying to figure it out from scratch every time.	Omitted	Omitted
22. If I have to, I can usually give reasons for my intuitions.	I	I
23. I prefer to follow my head rather than my heart.	(R) A	Analysis
24. I enjoy thinking in abstract terms.	HA	Omitted
25. I rarely trust my intuition in my area of expertise.	Omitted	Omitted
26. I try to keep in mind the big picture when working on a complex problem.	HB	HB
27. When I make intuitive decisions, I can usually explain the logic behind my decision.	I	I
28. It is foolish to base important decisions on feelings.	(R) A	Analysis
29. I am a "big picture" person.	HB	HB

In addition, questions used for a proxy of Analysis are indicated.

[a] Final 23-item scale

[b] Indicates the sole item included in both scales

Note: (R) = a reverse-worded   question;   I = Inferential;   A = Affective;   HA = Holistic   Abstract;   HB = Holistic Big Picture.

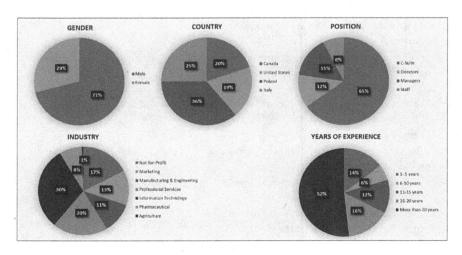

FIGURE 2.1

Demographic breakdown of the survey sample by gender, country, position, industry, and years of experience.

INTUITION STYLES: CORRELATIONS
BETWEEN THE CONSTRUCTS

In the initial paper, Pretz et al. (2014) reported a significant and material correlation between Affective Intuition and Holistic Abstract Intuition ($r = 0.298$, $p < .001$) but no other significantly correlated facets. As such, the other concepts were considered wholly independent constructs. The results of our research seem to contest this finding. While a material and significant correlation was found between Affective and Holistic Abstract Intuitions ($r = 0.361$, $p < .0005$), an even stronger and larger correlation was found between Holistic Big Picture and Inferential Intuition ($r = 0.472$, $p < .0005$). The full correlation matrix can be seen in Table 2.2.

These results seem to suggest the emergence of *intuition styles*. That is, while an individual has the capacity to independently leverage any of the four modes, there are patterns that emerge in each individual's propensity toward the different constructs. Those people who tend to leverage emotional characteristics in their judgments (i.e., those who use Affective Intuition) are largely the same types of people who will rely on a theoretical underpinning of the world (i.e., those who use Holistic Abstract Intuition). Similarly, individuals who are prone to rely on embedded analytical procedures are also likely to turn to a full-system view in synthesizing information (i.e., a combination of Inferential and Holistic Big Picture Intuition) (Figure 2.2).

Interestingly, the Analysis construct does not appear to be correlated strongly with the intuition constructs. There is a significant and somewhat material negative correlation between Inferential Intuition and Analysis ($r = -0.155$, $p = 0.042$), which gives credence to the proxy measure detecting what is meant by *analysis*. If Inferential Intuition represents embedded analytical procedures, which are no longer thought of as analysis as they have become second nature, then one should expect the direct reliance on Analysis to decrease, as the ability to rely on Inferential Intuition increases. It is important, however, to not read too deeply into the lack of observed correlation between the other metrics. This may very well be redundant mathematically, as the survey was administered based on an instrument that purported to measure only intuition, and the questions that make up the Analysis construct were found to not map onto the intuition constructs in the CFA. If we had observed significant and material correlations between the constructs, then these survey questions

	Holistic Abstract	Holistic Big Picture	Inferential	Affective	Analysis (Proxy)
Holistic Abstract		-0.003 (p=0.971)	-.152* (p=0.047)	.361** (p< .0005)	-0.038 (p=0.624)
Holistic Big Picture			.472** (p=0)	-0.056 (p=0.467)	-0.079 (p=0.305)
Inferential				.187* (p=0.014)	-.155* (p=0.042)
Affective					-0.07 (p=0.364)
Analysis (Proxy)					

FIGURE 2.2
Correlation matrix between the measured intuition constructs and analysis.

could have simply been incorporated into the four intuitive constructs. As such, to draw direct parallels between analysis and intuition, further research using different instruments is necessary.

Implications of the Findings

It is difficult to draw specific actionable insight from the appearance of intuitive styles alone. Being able to identify individuals who are likely to act as either Feeling Theorists or Big Picture Modelers may assist in the communication and distillation of ideas, directly extending the importance of recognizing intuition as a non-unitary construct. The assessment of ideas in a business often comes with varied opinions from the parties involved. Often, these differing ideas may simply be the result of everyone coming to an intuitive decision about the best path forward. It may then seem puzzling, if the naïve view of intuition is taken, as to how people come to differing views; these differences may be brushed off as the product of experience, positional bias, or ill-framed assumptions. While any of these factors may play a role, it also appears plausible that the differences emerge due to fundamentally different approaches to the problem. If it is possible to determine the style of intuition each thinker has, then it may also be possible to rationally understand the process that resulted in each decision. Understanding that emotions and theories may provide one answer, while a big-picture and analytical approach may provide another, seems to offer guidance for synthesizing solution differences and coming to the most efficient overall solution.

DYNAMIC DECISION MAKING: CHANGES THAT COME WITH EXPERIENCE

A primary goal of the study was to understand how the use of intuition changes with experience. A predictable, dynamic trend emerged across industries, positions, countries, and gender, which speaks to how experience influences one's judgment processes. Not only were there distinct intuition style preferences as employees gain experience, even when controlling for other covariates, but also there were patterns in individuals' reliance on Analysis. Early in their careers, employees exhibit a preference for acting as Feeling Theorists, relying on emotion and theory to guide their intuition. Additionally, those early in their career also tend to turn to analytical thought relatively frequently. As experience is gained, a gradual trade-off is observed. Workers move toward acting as Big Picture Modelers and let go of their reliance on Analysis. This pattern appears consistent with theory. Big Picture Modelers leverage embedded analytical procedures, which are learned through experience, explaining the lessened reliance on direct analysis and a heavier emphasis on Inferential Intuition. Further, as one gains relevant industry experience, abstract theories can be replaced by an intuited understanding of the complex structures that comprise the industry and position.

When controlling for position, inexperienced employees (with 1 to 5 years' experience) tend to trust in Affective and Holistic Abstract Intuition between 22% and 31% more than more experienced employees (with over 16 years of experience; $p < .05$). Similarly, inexperienced employees (with 1 to 5 years' experience) prefer Analysis between 17% to 27% more than employees with greater levels of experience (with over 11 years of experience; $p < .05$). Experienced employees (with over 20 years of work experience) report trusting their Inferential Intuition between 9% and 18% more so than less experienced employees (with both 1 to 5 and 6 to 10 years of experience; $p < .05$).

To further investigate this trend, the relative preferences for the differing styles of intuition were considered. To do so, the pairwise differences were computed and used in the same analysis. Again, controlling for position, relative preferences are also dynamic, changing as employees gain more experience. These trends fit exactly with the hypothesis of an intuition tradeoff. Experienced employees (with over 16 years of experience) have a significant, larger relative preference for Holistic Big Picture versus

Holistic Abstract Intuition, compared with their inexperienced colleagues (with 1 to 5 years of experience; $p < 0.05$). In other words, while all employees tend to prefer Holistic Big Picture to Holistic Abstract Intuition, the more experienced employees exhibit this behavior to a far greater extent (Figure 2.3). Similarly, all employees have a distinct preference for Inferential Intuition versus Affective Intuition. However, experienced employees (with over 11 years of experience) have a significant, larger relative preference versus their inexperienced colleagues (with 1 to 5 years of experience; $p < .01$). Finally, the analyses revealed that as employees gain experience, they tend to trust their Inferential and Holistic Big Picture Intuition versus Analysis to a greater extent than younger employees ($p < .05$) (Figures 2.4 through 2.6).

FIGURE 2.3
Relative preference for holistic big picture intuition versus holistic abstract intuition by years of experience.

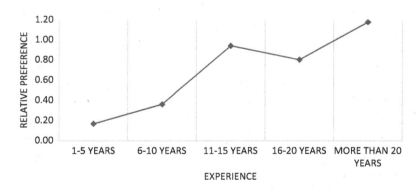

FIGURE 2.4
Relative preference for Inferential versus Affective Intuition by years of experience.

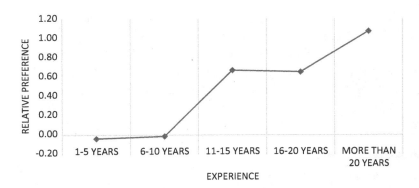

FIGURE 2.5
Relative preference for inferential intuition versus analysis by years of experience.

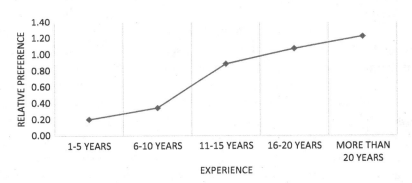

FIGURE 2.6
Relative preference for holistic big picture intuition versus analysis by years of experience.

Implications of the Findings

From this analysis, we see that younger employees and more experienced executives make decisions using intuition and data analysis quite differently. Executives are often accused of "ignoring the data," while analysts lack the big picture insight and experience to develop plausible models and interpret the results. Yet both bring important strengths to the decision-making process. For example, as Big Picture Modelers, executives can help give sense to the numbers and can assist with model development.

Given that the dynamic nature of these trade-offs is predictable, there appears to be an embedded learning process that may be leveraged once acknowledged. While the less senior employees eventually learn to adopt the executives' judgment processes, recognizing that the junior staff are using Analysis and acting as Feeling Theorists may give context to what they find. Instead of a conflictual "tug of war" between data analysis and intuition, changing the decision-making process to take advantage of the different types of intuition and analysis may result in more effective decisions. This may involve changes to how models are developed, how the results are interpreted, who is involved, and the iterative nature of the entire process.

THE CONUNDRUM: STYLES FOR MANAGERS VERSUS LEADERS

Given the dynamic changes observed as experience is gained, it is sensible to investigate whether there are any factors that influence these changes. Perhaps surprisingly, when comparing across different positions, and controlling for the number of years of experience, very few differences in intuitive processes emerge. It does not seem to matter whether someone is a C-suite executive or an entry-level staff member; if they have worked for the same amount of time, their decision-making profile likely will be similar. There is one significant difference from this trend, and it is one that may matter significantly when considering career progression. Managers report using Affective Intuition approximately 19% more than C-level executives, even when controlling for years of experience ($p < 0.01$). All other differences by position are not significant, except as proxies for the differences observed by experience.

Implication of the Findings

It may be the case that people who tend to heavily rely on Affective Intuition often find their way into management positions due to a natural alignment. It may also be the case that in order to act as a manager, it is necessary to rely on Affective Intuition. While it is not possible from this study to infer any causal mechanism, there are a number of possible conclusions to draw. Whether it is a static trait held by managers or a trait that the position naturally fosters, there are important considerations for individuals wishing to become managers and those wishing to progress beyond management.

If effective managers exhibit a propensity toward Affective Intuition due to, for example, focusing on the needs of those they manage, then this provides an actionable insight for those wishing to work in management. As employees gain experience, the trend is to shed their use of Affective Intuition. This is a trend that one may wish to avoid if the aspiration is to work as a manager.

However, this tendency potentially limits these managers in their career progression, if executives are expected to adopt a Big Picture Modeler approach to making decisions. Executives are sometimes required to make tough calls that are good for the business but may impact some stakeholders (e.g., employees and customers) negatively. Managers wishing to move into leadership positions may wish to better understand their intuition "style" and whether it may get in the way of effective leadership.

GLOBAL CONTEXT: THE SIMILARITIES ARE STRONGER THAN THE DIFFERENCES

The sample consisted mostly of individuals from Canada, the United States, Poland, and Italy. In testing for differences that may exist between nations, the overarching finding was that the countries were far more similar than they were different. It was only across Inferential Intuition and Analysis that significant differences were observed. Across the preferences for others, when holding the other covariates constant, country and continent did not play a significant role in explaining the differences. While it is important to not draw significant conclusions from a lack of trend, it is notable that, at least in this sample, culture seems to be less important than one might initially think.

When considering the preference for Analysis, employees in Italy reported using Analysis between 12% and 21% less than employees from Poland, Canada, and the United States (Figure 2.7; $p < 0.05$). The findings also suggest that employees from the United States tend to prefer Inferential Intuition 9% more than those from Italy (Figure 2.8; $p < 0.05$). These two differences together create a somewhat surprising result. While it is typically observed that analytical preference is traded off for Inferential Intuition, it is noted that for both attributes Italians report lesser reliance than the other countries. One may hypothesize that Italians are thus more likely to leverage Holistic Abstract and Affective Intuition, and while there is a slight preference for Affective Intuition, this is not significant. Moreover, the Italian sample is below average on Holistic Abstract as well, though this trend is also not statistically significant. As such, the

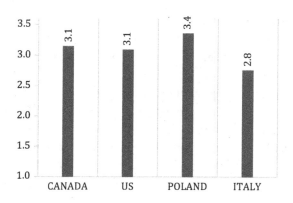

FIGURE 2.7
Preferences for analysis by country.

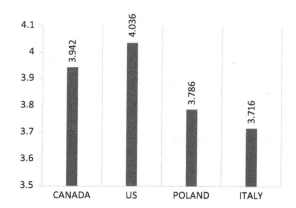

FIGURE 2.8
Preferences for holistic big picture intuition by country.

findings suggest that although national culture is not a dominant factor in the shaping of the intuition approaches taken by most individuals, it does affect subsets of individuals—for example, Italians who show a slight, non-significant preference for Affective Intuition.

Implications of the Findings

One important implication to draw is that, overall, differences across countries are not likely to be the most pronounced ones that are faced, in terms of judgment processes. The differences that do exist are almost certainly less material than the differences that exist, for instance, when comparing individuals with different years of experience. This is a positive finding for the encouragement of intra-team diversity, indicating that there may not be large hurdles to be overcome in decision making.

The differences that do exist need to be explored more deeply to determine their consequences. Given that those in our sample from Italy appear less likely to engage in analytical decision making, it may be the case that they are also less likely to be convinced by analytical arguments. Similarly, the trend toward Affective Intuition may suggest the tactics to best leverage in communication with Italian managers. The opposite advice could be given as an Italian engaged in work with, for instance, an American party; knowing that there will be a greater emphasis placed on Analysis provides insight into how to better engage the other party.

It is important to note, however, that while these differences have been observed, they are not necessarily an indication of best practice. The results of this study do not necessarily suggest that Italians are less impacted by analytical arguments or that Americans require them; it is merely an observation of what employees with these nationalities report using themselves.

INDUSTRY DIFFERENCES: MINOR IMPACTS ON INTUITIVE STYLE

Decision makers within certain industries tend to be stereotyped or caricatured. Asked to picture a tech CEO, a different image is almost certainly held from the image resulting from the request to picture an executive at a global investment bank. It is natural to assume, then, that these differences may present themselves in industry comparisons of decision-making styles. This is observed to an extent. Some industries exhibited tendencies away from certain forms of intuition, compared with the rest of the sample. While some of these differences are informative and actionable, and others are expected, the study did not find distinct clustering at the level that one may expect. For instance, it does not seem to be possible to accurately gauge whether an employee is an IT worker or a professional services worker simply based on their intuition profile.

The study found that those working in marketing firms reported using Analysis approximately 10% less than those outside of marketing ($p < .05$). When considering the manufacturing industry, firms here report using Holistic Abstract Intuition 21% less and Holistic Big Picture 11% less than firms outside of the manufacturing industry ($p < 0.05$). Firms in the IT industry reported using Inferential Intuition approximately 7% more than non-IT firms ($p < .01$), while firms outside of the IT industry reported using Affective Intuition approximately 10% more than IT firms ($p < .05$) (Figures 2.9 through 2.13).

Implications of the Findings

Some industries may have specific lessons to take from this analysis. It is interesting to observe that, for instance, the marketing industry does not report a high degree of reliance on analytics. It is possible that this

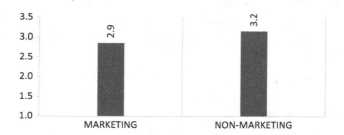

FIGURE 2.9

Preference for analysis: industry comparison (marketing vs. non-marketing).

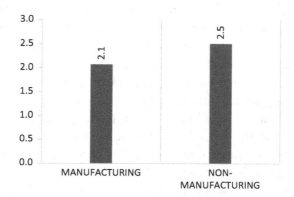

FIGURE 2.10

Preference for holistic abstract: industry comparison (manufacturing vs. non-manufacturing).

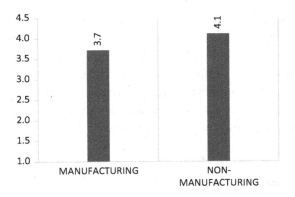

FIGURE 2.11
Preference for holistic big picture: industry comparison (manufacturing vs. non-manufacturing).

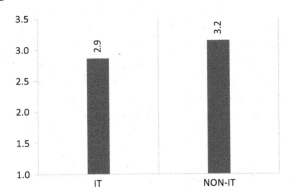

FIGURE 2.12
Preference for affective: industry comparison (IT vs. non-IT).

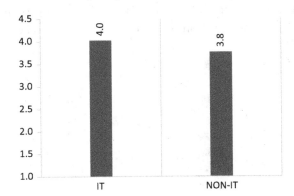

FIGURE 2.13
Preference for inferential: industry comparison (IT vs. non-IT).

limitation has been recognized, as there has seemingly been an increased push for customer analytics. Conversely, the IT industry's heavy analytical focus, even in intuitive terms, and lesser reliance on Affective Intuition may help explain some of the controversy surrounding the industry. There is a continued push for these companies to explore diversity and attempt to mitigate some of the social externalities that have arisen; this may be another axis along which progress can be measured.

GENDER: MINIMAL IMPACTS ON INTUITIVE STYLE

Comparisons by gender appear less material than comparisons by industry. The only significant difference found between males and females is that males report an approximately 8% higher propensity toward Holistic Big Picture Intuition compared with females ($p < .01$). While this difference remains when controlling for other factors, the exact mechanism driving this result is not clear. It may be the case, for instance, that men and women are intrinsically similar in their judgment profiles; it may also be the case, however, that it is through the work environments that conformance occurs (Figure 2.14).

Implications of the Findings

Gender differences appear to be less significant than some other factors we have examined. Similar to the observations made regarding countries, a key insight from this study appears to be that gender does not appear to be a driving force for intuition differences in the workplace. Instead, the

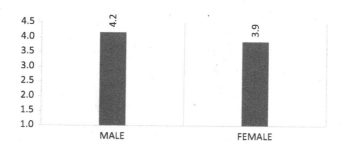

FIGURE 2.14
Preference for holistic big picture: gender comparison.

differences across gender appear to be minimal, which is good news for the integration of a gender-diverse team.

RESEARCH SUMMARY

This survey, and the exploratory analysis that followed, unearthed a number of interesting trends regarding the use of intuition and analysis in decision-making processes. The major contributions of the study are as follows:

- Styles of intuition exist: (1) *Feeling Theorists*: employees who prefer to trust their emotional intelligence (Affective Intuition) and also prefer to use "abstract" theory in their decision making. (2) *Big-Picture Modelers:* employees who prefer using Inferential Intuition and also prefer "seeing the big picture."
- The use of and preference for different styles of intuition as well as analysis are dynamic, changing as employees gain more experience. Importantly, this change with experience is predictable and stable across different industries, genders, and countries, and this trend remains even when controlling for years of experience.
- The less experienced the employee, the more likely will the employee act as Feeling Theorists. As experienced is gained, the tendency is to become a Big Picture Modeler, both in absolute and relative terms. This trend is present even when controlling for other covariates.
- The less experienced the employee, the more likely will the employee leverage direct analysis. As experience is gained, there is a corresponding decrease in the reliance on Analysis. This trend is present even when controlling for other covariates.
- Managers have a higher preference for Affective Intuition than C-level executives, regardless of their years of experience. This suggests that certain traits may overcome the dynamism, allowing intuition styles to remain static and trait-like, differentiating managers from leaders.
- Moderate country-level differences in preferences for intuition and analysis exist. Employees from Poland, Canada, and the United States prefer Analysis more than employees in Italy, while employees from the United States prefer using Inferential Intuition more than their Italian counterparts. This implies that there may be some cross-cultural context factors to consider in workplaces, but that on the

whole, nationality is not of primary importance in a decision-making context.

- Moderate industry differences exist in preferences for the decision-making process. The marketing industry executive is less likely to leverage analytics; the manufacturing industry supervisor is less likely to use Holistic (both Abstract and Big Picture) Intuition; and the IT industry manager is less likely to use Affective Intuition and more likely to use Inferential Intuition. These differences do suggest some industry-specific lessons to learn; for instance, a heavier focus on Analysis in marketing may yield positive benefit. They also suggest that, for the most part, industries are more similar than different in their use of intuition in decision making—at least given this sample.
- Gender appears to be a relatively unimportant variable in terms of intuition and analysis. Males are slightly more inclined toward Holistic Big Picture Intuition.

ACKNOWLEDGMENTS

We would like to thank the team of researchers, including Jay Liebowitz, Joanna Paliszkiewicz, and Fabio Babiloni, who contributed toward this research. We also thank Fulbright Canada, the CIES, and the Smith School of Business, Queen's University for supporting this research.

REFERENCES

O'Connor, B. P. (2017). Package "paramap". R package Version 1.4. Retrieved from https://people.ok.ubc.ca/brioconn/nfactors/paramap.pdf.

Pretz, J. E., and Totz, K. S. (2007). Measuring individual differences in affective, heuristic, and holistic intuition. *Personality and Individual Differences, 43* 1247–1257.

Pretz, J., Brookings, J., Carlson, L., Humbert, T., Roy, M., Jones, M., and Memmert, D. (2014). Development and validation of a new measure of intuition: the types of intuition Scale. *Journal of Behavioral Decision Making, 27*(5), 454–467.

Velicer, W. (1976). Determining the number of components from the matrix of partial correlations. *Psychometrika, 3* (41), 321–337. doi:10.1007/bf02293557.

3

It Is a Matter of Heart: C-Level Executives Detect Their Heartbeat Better than Other Company Members Do

Written by: Fabio Babiloni, Enrica Modica, and Patrizia Cherubino

CONTENTS

INTRODUCTION

Interoception awareness (IA) is commonly defined as the self-appreciation of the information flow from the internal organs (Critchley & Garfinkel, 2017). IA has been posed as the center of the theory of emotion, suggesting that emotional feelings are heavily related to central representation and the perception of changes in bodily physiology (Lange & James, 1967). As a consequence, persons who are more sensitive to bodily responses experience emotions with increased strength (Wiens, Mezzacappa, and Katkin, 2000).

In fact, scientists have underlined how and when the perception of internal body signals could enhance and drive cognition and decision making (Dunn et al., 2010) as well as improving memory (Garfinkel et al., 2013; Werner et al., 2009, 2010). Although there are several methods of measuring IA (e.g., respiration rates, heartbeat detection, etc.), heartbeat detection tasks dominate the scientific literature. In fact, heartbeats are easy to measure, are frequent internal events in the body, and can be sufficiently discriminated by the participants in an experimental study.

Interest in measuring IA using the heartbeat relies on the fact that several studies have suggested a strict correlation between IA capabilities and a particular class of subjects that generate accurate decision making. In particular, it has been suggested that higher scores on heartbeat detection predict superior performance in some laboratory gambling tasks (Crone et al., 2004; Dunn et al., 2010; Werner et al., 2009) or success on the stock exchange in London (Kandasamy et al., 2016) as well as avoiding losses (Sokol-Hessner et al., 2015).

It is well known that C-level executives have to deal in their daily activities with a continuous decision-making exercises. However, a surprisingly large amount of them when asked through questionnaires and interviews reported that most of their decision making was provided not only on the basis of quantitative data but also on their "intuition" (see other chapters in this book for references). Thus, it is of interest to understand whether such intuition could be linked to a clear increase in IA when compared with other employees not directly involved at the C level. For this reason, a study of C-level executives and their IA capabilities by heartbeat detection was conducted to compare with those offered by a group of working colleagues not at the C level. The experimental question was whether there was a significant difference in IA ability between the two analyzed groups. In addition, it was also analyzed whether the IA ability of C-level executives had some correlation with the explicit answers to the intuition questionnaire by Pretz et al. (2014).

METHODS

Subjects

Two groups of subjects were investigated. The first group (Executive) was comprised of 15 executives (balanced for gender) having C-level duties in small, medium, and big Italian companies. The second group (Control) was formed by 30 employees recruited at different large Italian financial and telecommunication companies with intermediate levels of responsibility (i.e., not C level but with a least 15 years of work inside the company). All subjects in both groups obtained their masters degrees in business, communication, and/or finance.

All subjects participated voluntarily in the study and each one of them gave written informed consent in accordance with the Declaration of Helsinki of 1975, as revised in 2000. The research project related to this study has received the approval of the proper ethical committee of the University of Rome Sapienza, Department of Physiology and Pharmacology. Measurements of the subjects were taken in the workplace on different days during normal working hours.

Experimental Tasks

Each participant was seated comfortably alone in a quiet meeting room inside the building of the company, with two experimenters and the related measurement device. Two tasks were required of all the participants.

The first one was a heartbeat counting task (Kandasamy et al., 2016) followed by a judgment by the participant about the quality of this self-estimation. The second one was a time lapse detection task (Kandasamy et al., 2016), employed to estimate the capability of the participant to internally count the duration of randomized time intervals proposed by the experimenter.

At the end of the measurements, all the participants were invited to compile a questionnaire (Pretz et al., 2014) about their style of decision making.

Heartbeat Task

Before the heartbeat counting task, each participant received the following instruction from the experimenter: *Without manually checking any part of your body, you have to silently count each heartbeat you feel from the time you hear "start" from me to when you hear "stop," always by me.*

The task was performed five times, using different time durations of 25, 30, 35, 45 and 50 seconds, randomized in their occurrence across each participant. At the end of each round of the heartbeat detection task, the participants have to score their confidence in the estimate provided on a seven-point Likert scale, from 1 ("Total guess/No heartbeat awareness") to 7 ("Complete confidence/Full perception of heartbeat").

Time Lapse Task

After the heartbeat counting task, each participant during the time lapse task received the following instructions from the experimenter: *Without speaking, you have to silently count the time passing in seconds from when you hear "start" from me to when you hear "stop," always by me.*

After each silent counting, participants were required to share their detection of time with the experimenter. The task was performed three times, using three different time durations of 19, 37, and 49 seconds, randomized across each participant.

Electrocardiographic Measurements

All the participants in the experimental study took their electrocardiographic measurements by using a portable Nexus-10 system (Mind Media, the Netherlands). In particular, electrocardiogram (EKG) signals were continuously acquired for the entire duration of the experiment by using two electrodes across the participants' fingers. The EKG signals were filtered and processed offline with the use of in-house MATLAB software for the detection of the heartbeats of the participants.

Estimation of Participant's Accuracy of the Detection of Their Heartbeats and Time Lapses

In agreement with the previously published procedures (Kandasamy et al., 2016), the heartbeat detection accuracy score was expressed as a percentage using the following equation:

Heartbeat score (HBS)

$$= \left(\left[1 - \left[\frac{\text{absolute value of} \left(\text{actual heartbeat} - \text{estimated heartbeat} \right)}{\div \text{ mean of} \left(\text{actual } n + \text{estimated } n \right) \text{heartbeat}} \right] \right] \right) \times 100.$$

The same methodological approach was performed by estimating the Time Difference Score (TDS) as follows:

Time difference score (TDS)

$$= \left(\left[1 - \left[\frac{\text{absolute value of} \left(\text{actual time lapse} - \text{estimated time lapse} \right)}{\div \text{ mean of} \left(\text{actual } n + \text{estimated } n \right) \text{ time lapse}} \right] \right] \right) \times 100.$$

HBS and TDS were estimated for each subject across the five repetitions for heartbeat detection and three repetitions for time lapse estimation.

Statistical Analysis

In order to confirm the HBS index did not change across the experiment, a repeated measure ANOVA was performed with the factor TRIAL (with five levels, related to heartbeat detection with different time lapses employed) and the between-factor GROUP (two levels: Executive, Control).

ANOVA with repeated measures was also performed to confirm the TDS index did not change between Executives and Control, with the factors GROUP (two levels: Executive, Control) and TIMELAPSE (with three levels related to the different time lapses submitted to the subjects).

Correlations between the HBS and TDS indexes and the age of operation of the participants in the Executive group were measured by Pearson correlation, with a predefined significance level of $p = .05$.

Correlations between the values of the HBS index for the Executive group and the different scores obtained by the same participants in the questionnaire on different forms of intuition were measured with a Pearson correlation coefficient, with a significance level of $p = .05$.

RESULTS

In the following, we report the results obtained from the analysis of the HBS and TDS indexes for the analyzed Executive and Control groups. No significant correlations between the HBS indexes and the different dimensions of the questionnaire on the style of decision making were found for the participants in the Executive group.

Executives Have Enhanced Interoceptive Ability Compared with Controls

Results from the ANOVA showed the statistical significance of the factor GROUP, with $F(1.43) = 9.74$ associated with $p = .0032$, while the factor TRIAL returned a value of $F(4.172) = 2.02$ with $p = .093$ and the factor GROUP × TRIAL had a value of $F(4172) = 1.23$ with $p = .29$.

Figure 3.1 shows the situation for the average values of the HBS index between the Executive and Control groups, where it is possible to observe the prevalence of the heartbeat detection from the Executive group compared with the Control group.

In addition, as a consequence of the ANOVA results for the factor TRIAL, it is possible to state that the HBS index does not change significantly during the experiment in both groups ($F[4.172] = 2.02$; $p = .093$). Figure 3.2 shows the relatively constant average values of the HBS index across the five trials of the heartbeat detection experiment in the two groups examined.

Executives' HBS Indices Are Inversely Correlated with Their Years of Operation

The Pearson correlation between the HBS index and the years of operation of the participants in the Executive group is statistically significant ($r^2 = .292$, with an associated $p = .0377$). The slope of this correlation is negative; that is, the more years of operation, the lower the value of the

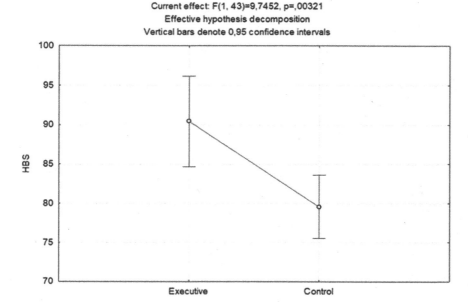

FIGURE 3.1

Average values of the HBS index for the Executive and Control groups. Differences between the HBS values of the two groups are statistically significant (p = .0032).

HBS index among participants in the Executive group. Figure 3.3 shows the correlation between the HBS index and the years of operation of the participants in the Executive group.

Both Executive and Control Groups Have Similar Perceptions of Time Lapses

The ANOVA performed on the values of the TDS index suggested that not all the factors considered (GROUP and TIMELAPSE) are statistically significant. In particular, the ANOVA on the TDS index showed that the factor GROUP exhibited a value of $F(1.43) = .037$, with an associated $p = .84$. In addition, the factor TIMELAPSE was not statistically significant, with $F(2.86) = .48$ and an associated $p = .61$, while the factor GROUP×TIMELAPSE is also not statistically significant, with $F(2.86) = .32$ and an associated $p = .72$.

In addition, it was also analyzed whether the participants in the Executive group had an efficiency in time perception that was correlated with the years of operation. This was performed by a Pearson correlation

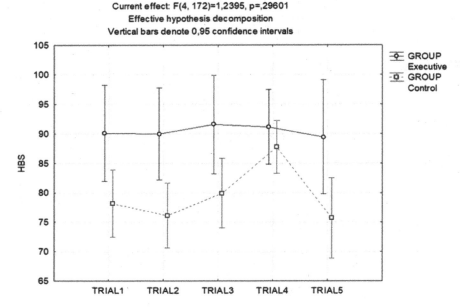

FIGURE 3.2

Values of the HBS index across the two groups during the repetition of the heartbeat detection experiment. Statistical analysis suggests the equivalence of the accuracy of the heartbeat detection across the different trials performed.

between the TDS values and the years of service at the C level. The results suggest that the correlation coefficient between the TDS and the years of operation of the participants in the Executive group was not statistically significant ($r^2 = .19$, with an associated $p = .11$).

DISCUSSION

The results obtained in this study will be discussed in the following subsections, each one relative to the particular result obtained in the study.

Executives Have Enhanced Interoceptive Ability Compared with Controls

Results obtained from EKG recordings clearly support the fact that the participants in the Executive group show an interoceptive ability, as

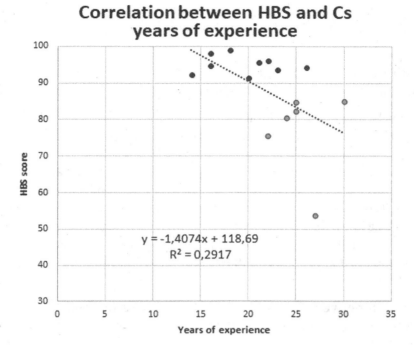

FIGURE 3.3

Correlation between the HBS index and the years of experience of the participants in the Executive group. The light-gray dots are relative to the participant in the Executive group that has an HBS lower than the median of the group. Instead, the black dots are relative to participants whose HBS is higher than the median.

indexed by the HBS, that exceeds the values obtained by the Control group in a way not due to chance alone.

This ability, often called *interoceptive awareness* in the literature, has been highlighted in other professional groups that use intuition-based decision making, such as stock traders in London (Kandasamy et al., 2016). Previous studies have reported lower values on the HBS index for both stock traders and controls than those observed here for the Executive and Control groups, in a range of about 10%. In fact, Kandasamy et al. (2016) reported average HBS index values of 78.2% for traders and 66.9% for control subjects, which are lower than those observed here. While this 10% discrepancy could be due to the limited sample sizes used in both studies, the difference observed here in the HBS index between Executive and Control groups was instead statistically relevant.

A better IA was also recently demonstrated in different groups of subjects who presented a superior ability to emphatically understand others

(Shah et al., 2017) or who have specific visuo-motor abilities (Christensen et al., 2018) compared with a control group.

Both Executive and Control Groups Have Similar Perceptions of Time Lapses

It may be argued that the differences in interoception as indexed by HBS could be due to the use of particular counting strategies by the two groups rather than the genuine detection of internal bodily activity. However, this hypothesis is not supported by results relating to the statistical similarity of the time lapse detection tasks for both Executive and Control groups. Thus, it could be concluded that the heartbeat detection task was performed by taking into account interoceptive information instead of counting time lapses internally. In fact, both groups were able to precisely address such issues without any significant differences between them.

Relation between Executives' Years of Operation and Their Interoceptive Sensitivity and Time Lapse Estimation Ability

In this experiment, information about executives' years of operation at the C level was collected and correlated with the indices related to interoceptive sensitivity (HBS) and the ability to detect time lapses (TDS).

Interestingly, participants in the Executive group have shown a significant inverse correlation between the values of the HBS index and the age of operation, in the sense that the HBS index decreases significantly for executives who have more than 25 years of experience. In fact, the difference in the years of operation between the two groups of executives characterized by high and low HBS is statistically relevant. In particular, the maximum of the HBS index is for executives with between 11 and 25 years of operation. Executives with more than 25 years of operation have a lower HBS index. These results are consistent with previous observations suggesting interoceptive ability decreases with age (Khalsa et al., 2009; Murphy et al., 2018).

The evaluation of time lapses by TDS index for the participants in the Executive group was instead not correlated with the age of operation, and it is also equivalent to the Control group. This means that time perception plays no role in the explanation of the different results for the HBS index in the two populations examined.

Can IA Be Trained?

Published literature supports the idea that high IA is correlated with a more precise appreciation of emotional feelings (Pollatos et al., 2007) and cognitive decision making, and also physical activity (Werner et al., 2009). Furthermore, it has also been pointed out that IA is linked to cognitive affective processing in adaptive intuitive decision making (Dunn et al., 2010) and in decision making during game theory (Lenggenhager et al., 2013). Thus, the results highlight that interoception plays a role beyond the field of pure emotion, entering the decision-making context.

Since IA is important in decision making and is significantly increased in senior C-level executives and in successful stock exchange traders, it is a quality of potential interest for future young or candidate executives. An immediate question then arises: Can this awareness be trained? The answer from the scientific literature isn't quite clear. In fact, objective interoceptive accuracy in heartbeat detection was not previously associated with an increase in body awareness after a period of training and practicing (Schaefer et al., 2014). Furthermore, it was also demonstrated that experienced meditators presented similar heartbeat detection accuracy to a control group (Khalsa et al., 2008). Interestingly, context meditators consistently rated their accuracy higher when compared with controls (Khalsa et al., 2008).

On the other hand, a recent paper presented interesting results related to a statistically significant increase in IA after attending three meditation courses, each of 3 months, for a total of 9 months of practice in a consistent cohort of participants (Bornemann and Singer, 2017).

Thus, by undergoing intensive meditation practice, an increase in IA can be obtained. This last aspect could underline the interest in meditation training arising recently among managers in the business world. However, it must be underlined that the actual literature shows evidence of an increase in interoception accuracy and sensitivity only after 9 months of regular practice, which is quite far from the standard 8 weeks of courses usually offered to managers.

Future Direction of Research

To date, the research has pointed out that IA is linked to a better perception of emotions in particular classes of participants, who showed improved quality in their decision making with respect to controls. However, this correlation between IA and improved decision making is somewhat indirect: people who had a very good business career (stock exchange traders, C-level managers) presented high levels of IA. Future research needs to better address the issue of how emotions, decision making, and IA are connected, by manipulating for instance the emotional mood of the participants during their (possibly financial) decision making. In particular, it would be interesting to evaluate whether a person who has high IA could be more sensitive than others to emotionally induced stressors (e.g. fear, anger, sadness) and how this could impair (or improve) the efficiency of their decision making. The results of these future studies could be quite informative for practical applications, since it also well known that a manager's decision making occurs in daily life under a variety of emotional moods. If the results indicate that, under a variety of different emotional moods and stressors, the quality of decision making is improved in people who have IA compared with controls, there will be a clear indication that good decision making is "a matter of heart."

REFERENCES

Bornemann, B., and Singer, T. (2017). Taking time to feel our body: Steady increases in heartbeat perception accuracy and decreases in alexithymia over 9 months of contemplative mental training. *Psychophysiology*, 54, 469–482. http://dx.doi.org/10.1111/psyp.12790.

Christensen, J. F., Gaigg, S. B., and Calvo-Merino, B. (2018). I can feel my heartbeat: Dancers have increased interoceptive accuracy. *Psychophysiology*, 55(4). doi: 10.1111/psyp.13008. Epub September 21, 2017.

Crone, E. A., Somsen, R. J., Beek, B. V., and Van Der Molen, M. W. (2004). Heart rate and skin conductance analysis of antecedents and consequences of decision making. *Psychophysiology*, 41, 531–540.

Critchley, H., and Garfinkel, S. (2017). Interoception and emotion. *Current Opinion in Psychology*, 17, 7–14.

Dunn, B. D., Galton, H. C., Morgan, R., Evans, D., Oliver, C., Meyer, M., Cusack, R., Lawrence, A. D., & Dalgleish, T. (2010). Listening to your heart: How interoception shapes emotion experience and intuitive decision making. *Psychological Science*, 21(12), 1835–1844.

Garfinkel, S. N., Barrett, A. B., Minati, L., Dolan, R. J., Seth, A. K., and Critchley, H. D. (2013). What the heart forgets: Cardiac timing influences memory for words and is modulated by metacognition and interoceptive sensitivity. *Psychophysiology*, 50(6), 505–512. http://dx.doi.org/10.1111/Psyp.12039.

Kandasamy, N., Garfinkel, S., Page, L., Hardy, B., Critchley, H., Gurnell, M., and Coates, J. (2016). Gut feelings and the market: Interoceptive ability predicts survival on a London trading floor. *Scientic Report*, 6, 32986.

Khalsa, S. S., Rudrauf, D., Damasio, A. R., Davidson, R. J., Lutz, A., and Tranel, D. (2008). Interoceptive awareness in experienced meditators. *Psychophysiology*, 45(4), 671–677. http://dx.doi.org/10.1111/j.1469-8986.2008.00666.x.

Khalsa, S. S., Rudrauf, D., and Tranel, D. (2009). Interoceptive awareness declines with age. *Psychophysiology*, 46, 1130–1136.

Lange, C. G., and James, W. (1967). *The Emotions*. New York: Hafner (edited by Knight Dunlap, reprinted).

Lenggenhager, B., Azevedo, R. T., Mancini, A., and Aglioti, S. M. (2013). Listening to your heart and feeling yourself: Effects of exposure to interoceptive signals during the ultimatum game. *Experimental Brain Research*, 230(2), 233–241.

Łukowska, M.., Sznajder, M., and Wierzchoń M. (2018). Error-related cardiac response as information for visibility judgements. *Scientific Report*, 8(1), 1131. doi: 10.1038/s41598-018-19144-0.

Murphy, J., Geary, H., Millgate, E., Catmur, C., and Bird, G. (2018). Direct and indirect effects of age on interoceptive accuracy and awareness across the adult lifespan. *Psychonomic Bulletin & Review*, 25(3), 1193–1202. doi: 10.3758/s13423-017-1339-z.

Pollatos, O., Gramann, K., and Schandry, R. (2007). Neural systems connecting interoceptive awareness and feelings. *Human Brain Mapping*, 28(1), 9–18. http://dx.doi.org/10.1002/Hbm.20258.

Pretz, J., Brookings, J., Carlson, L., Humbert, T. K., Roy M., Jones M., and Memmert, D. (2014). Development and validation of a new measure of intuition: The types of intuition scale. *Journal of Behavioral Decision Making*, 27(5), 454–467. doi:10.1002/bdm.1820.

Shah, P., Catmur, C., and Bird, G. (2017). From heart to mind: Linking interoception, emotion, and theory of mind. *Cortex*, 93, 220–223. http://dx.doi.org/10.1016/j.cortex.2017.02.010.

Schaefer, M., Egloff, B., Gerlach, A. L., and Witthoft, M. (2014). Improving heartbeat perception in patients with medically unexplained symptoms reduces symptom distress. *Biological Psychology*, 101, 69–76.

Sokol-Hessner, P., Hartley, C. A., Hamilton, J. R. and Phelps, E. A. (2015). Interoceptive ability predicts aversion to losses. *Cognition and Emotion*, 29, 695–701.

Werner, N. S., Jung, K., Duschek, S., and Schandry, R. (2009). Enhanced cardiac perception is associated with benefits in decision-making. *Psychophysiology*, 46, 1123–1129.

Werner, N. S., Peres, I., Duschek, S., and Schandry, R. (2010). Implicit memory for emotional words is modulated by cardiac perception. *Biological Psychology*, 85(3), 370–376. http://dx.doi.org/10.1016/j.biopsycho.2010.08.008.

Wiens, S., Mezzacappa, E. S., and Katkin, E. S. (2000). Heartbeat detection and the experience of emotions. *Cognition and Emotion*, 14(3), 417–427.

4

Trust and Intuition in Management

Written by: Joanna Paliszkiewicz

CONTENTS

INTRODUCTION

Trust and intuition have increasingly fascinated researchers (Liebowitz, Paliszkiewicz, and Gołuchowski, 2018). Trust and intuition are important in responses to crises in an increasingly uncertain and changing world. Their role is especially significant in the era of big data with access to large volumes of information, where decisions must be made rapidly. Researchers look at trust and intuition from many perspectives and in many areas of knowledge—for example, psychology, sociology, and management (Mayer, Davis, and Schoorman, 1995; Gilovich, Griffin, and Kahneman, 2002; Hogarth, 2001; Klein, 1998; Sztompka, 1999; Myers, 2002; Gigerenzer, 2007; Plessner, Betsch, and Betsch, 2008; Sadler-Smith, 2008; Sprenger, 2009; Paliszkiewicz et al., 2014, Paliszkiewicz and Koohang, 2016).

The aim of this chapter is to present the role of trust and intuition in management. First, the conceptual background of trust and intuition is discussed. Second, concrete suggestions are made on what leaders can do to build trust and enhance intuition in organizations. Third, recommendations for future research are presented.

CONCEPTUAL BACKGROUND OF TRUST

Streams of research on the topic of trust can be found in different disciplines—for example, psychology (Rotter, 1971; Johnson-George and Swap, 1982; Simpson, 2007), sociology (Lewis and Weigert, 1985; Yamagishi Cook, and Watabe, 1998; Sztompka, 1999; Molm, Takahashi, and Peterson, 2000), and management (Colquitt, Scott, and LePine, 2007; Dirks and Ferrin, 2002; Kramer and Lewicki, 2010; Lewicki, Tomlinson, and Gillespie, 2006; Mayer, Davis, and Schoorman, 1995; McAllister, 1995; McEvily, 2011; Paliszkiewicz et al., 2014; Paliszkiewicz and Koohang, 2016). Most researchers see trust as a personality trait, a belief, and a positive expectation.

Psychologists describe trust as a personality trait that reflects the general expectations of the trustworthiness of others (Wrightsman, 1966; Rotter, 1971). According to Luhmann (1979), trust has to be learned, but according to Gibb (1978), trust is instinctive and labeled as a feeling very close to love. Trust can be considered a mix of feeling and rational thinking.

People vary in terms of when and how much they are willing to trust (Worchel, 1979; Das and Teng, 2004). Trust as a trait is responsible for the general expectations of the trustworthiness of others (Wrightsman, 1966; Rotter, 1967). These expectations depend on previous experiences and estimations of the probability that those trusted will reciprocate the trust (Tyler and Kramer, 1996). The expectations stem from cultural backgrounds as well (Mayer, Davis, and Schoorman, 1995). McKnight, Cummings, and Chervany (1998) identified two types of disposition to trust: *faith in humanity* and a *trusting stance*. Faith in humanity refers to the belief that others are reliable, whereas having a trusting stance means that people believe they will obtain better interpersonal outcomes by dealing with others as though they are reliable, regardless of whether they are reliable or not. Trust is also very important in sociology; it is the essential ingredient in the initiation and maintenance of stable social relations (Blau, 1964). People would have no occasion to trust apart from their relationships with others (Lewis and Weigert, 1985). Luhmann (1979) argues that trust reduces social complexity, as it simplifies life by taking risks. Barber (1983) identifies three kinds of expectations in relation to trust: (1) an expectation of the persistence and fulfillment of the natural and social order, (2) an expectation of competent role performance, and (3) an expectation that partners in interactions will carry out their fiduciary responsibilities. The three presented perspectives on trust highlight trustors' beliefs that the trustees are honest and good despite their ability to betray, so the risk is acceptable. Risk is indispensable in the cultivation of trust because trust would not be necessary if actions could be pursued with absolute certainty (Lewis and Weigert, 1985). In Bachmann's (1998) view, trust is necessary for situations in which trustors have partial information about the factors that will possibly influence trustees' future behavior.

Trust has emerged as a central construct in a wide range of management studies, including those focusing on performance (Kramer, 1999; Lewicki, Wiethoff, and Tomlinson, 2005; Mayer, Davis, and Schoorman, 1995; McAllister, 1995; Colquitt, Scott, and LePine, 2007). The concept of trust in management indicates employee faith in an organization's goal attainment, organizational leaders, and the belief that organizational action will be beneficial for employees (Kim and Mauborgne, 1997). Research has revealed that employees' trust is linked to their working attitudes and behaviors (Aryee, Budhwar, and Chen, 2002; Atuahene-Gima and Li, 2002; Dirks and Ferrin, 2002). Trust is important in an organization because of its influence on successful cooperation and efficiency in organizations (Lewis and Weigert, 1985; McAllister, 1995; Nooteboom, 2002; Rousseau et al., 1998; Zand, 1997). Trust is essential for innovative work and learning (Bartsch, Ebers, and Maurer, 2013; Jones and George, 1998); it binds friendships (Gibbons, 2004) and facilitates negotiations (Olekalns and Smith, 2005).

In summary, trust is based on personality traits. People vary in terms of whom they are able to trust. Trust is interpersonal and appears between concrete individuals. Trust depends on the behavior of other people. Trust cannot be compulsory. It springs from choice. Trust cannot be forced on anybody. Trust is difficult to build and maintain, but it is very easy to destroy. Trust appears to be a factor that, if reciprocated, strengthens relationships over time and is not consumed, like other forms of capital.

CONCEPTUAL BACKGROUND OF INTUITION

People often rely on intuition, which is the psychological function that allows knowing to occur by means of transmitting perceptions in an unconscious way (Jung, 1924). Philosophers often conceive of intuition as competence (Betsch, 2008). Adopting the philosophical approach, in the *Encyclopædia Britannica* we can find the definition of intuition as "the power of obtaining knowledge that cannot be acquired either by inference or observation, by reason or experience." Intuition enables individuals to "grasp the meaning, significance, or structure of a problem without explicit reliance on analytical apparatus"; hence, it can "synthesize disparate ideas, achieving serendipity as it senses combinations which did not appear to be related in the past" (Isaack, 1978, p. 919).

According to Betsch (2008, p. 4):

> Intuition is a process of thinking. The input to this process is mostly provided by knowledge stored in long-term memory that has been primarily acquired via associative learning. The input is processed automatically and without conscious awareness. The output of the process is a feeling that can serve as a basis for judgments and decisions.

Shapiro and Spence (1997, p. 64) define intuition as "a nonconscious, holistic processing mode in which judgments are made with no awareness of the rules or knowledge used, for interference and can feel right despite one's inability to articulate the reason." In the literature, there are many propositions for defining the word *intuition*. Several examples are as follows:

- An immediate awareness of the subject, of some particular entity, without such aid from the senses or from reason as would account for that awareness (Wild, 1938, p. 226).
- The process of reaching a conclusion on the basis of little information, normally reached on the basis of significantly more information (Westcott and Ranzoni, 1963, p. 595).
- A preliminary perception of coherence (pattern, meaning, structure) that is at first not consciously represented but that nevertheless guides thought and inquiry toward a hunch or hypothesis about the nature of the coherence in question (Bowers et al., 1990, p. 74).
- It is as a brain skill (Lank and Lank, 1995).
- A feeling of knowing with certitude on the basis of inadequate information and without conscious awareness of rational thinking (Shirley and Langan-Fox, 1996, p. 564).
- A cognitive conclusion based on a decision maker's previous experiences and emotional inputs (Burke and Miller, 1999, p. 92).
- A tacit form of knowledge that orients decision making in a promising direction (Policastro, 1999, p. 89).

- A perceptual process, constructed through a mainly subconscious act of linking disparate elements of information (Raidl and Lubart, 2000–2001, p. 219).
- Thoughts that are reached with little apparent effort, and typically without conscious awareness; they involve little or no conscious deliberation (Hogarth, 2001, p. 14).
- The capacity for direct, immediate knowledge prior to rational analysis (Myers, 2002, pp. 128–129).
- Thoughts and preferences that come to mind quickly and without much reflection (Kahneman, 2003, p. 697).
- It is affectively charged judgment that arises through rapid, non-conscious, and holistic associations (Dane and Pratt, 2007).

The characteristics of intuition have been outlined in the literature. For example, Badke-Schaub and Eris (2014) and Shapiro and Spence (1997) proposed the following features of intuition:

- The source of intuition occurs at a non-conscious level.
- Intuition is fast.
- Intuition uses multi-sensorial stimuli.
- Intuition is accompanied by affect/feelings/emotions.
- Intuition develops with experience.
- Intuition involves a holistic interpretation of information.
- Intuition can stimulate creative solutions.

In management, intuition has been posited to help guide a wide range of critical decisions (Dane and Pratt, 2007). Research suggests that intuition may be integral to successfully completing tasks that involve high complexity and short time horizons, such as corporate planning, stock analysis, and performance appraisal (Hayashi, 2001; Isenberg, 1984; Shirley and Langan-Fox, 1996).

BUILDING TRUST AND ENHANCING INTUITION

The process of building and developing trust in an organization is very relevant and it is always related to a certain amount of risk. It is especially important in the situation when the trustor depends on the trustee's future actions to achieve his/her own goals and objectives (Lane, 1998). Building trust is an interactive process that involves (at least) two individuals learning about each other's trustworthiness (Zand, 1972; Zucker et al., 1996) under a given contextual and/or structural organizational setting. Trust is built up gradually, reinforced by previous trusting behavior and previous positive experiences (Zand, 1972; McAllister, 1995; Lewicki and Bunker, 1996).

People's readiness to trust others grows with their ability to trust themselves. If people believe that they are dependable and reliable, and see themselves as trustworthy, others will be more willing to put their trust in them. In order to place trust in a partner, one must first establish his/her trustworthiness, credibility, and reputation (Covey, 2009). Based on the characterization of Mayer, Davis, and Schoorman (1995), people are assessed to be trustworthy when they

- Have the required skills, competencies, and characteristics that enable them to exert influence within a specific domain—a description of the *ability* or *competence* criterion
- Are believed to do good to trustors, setting aside an egocentric motive—thereby meeting the *benevolence* criterion
- Are perceived to adhere to a set of principles that trustors consider acceptable—a definition of *integrity*

Interpersonal trust or distrust occurs automatically before any conscious or deliberative thinking (Huang and Murnighan, 2010). There are a number of factors that influence initial decisions to build trust. They are

- Personality (Evans and Revelle, 2008)
- Current moods and emotions (Dunn and Schweitzer, 2005; Lount, 2010)
- An individual's physical appearance (Krumhuber et al., 2007)
- Dispositions to trust (Righetti and Finkenauer, 2011)

Many researchers have confirmed that intuitive reasoning plays a significant role in social interactions. For example, intuitive processes have been found to play essential roles in terms of the communication of social perceptions (Andersen, Reznik, and Manzella, 1996), judgments and interpersonal evaluations (Glaser and Banaji, 1999), persuasion (Chaiken, Liberman, and Eagly, 1989), and quick judgments about those we have just met (Andersen et al., 1995).

Trust is always fragile. A minor signal of distrust can freeze the attempt to develop a relationship. Facial clues can provide information about whether an individual can be approached, avoided, trusted, or distrusted (Berry and Brownlow, 1989; Van't Wout and Sanfey, 2008). According to Scharlemann et al. (2001), smiling has been found to influence the decision to trust in economic games. Willis and Todorov (2006) insisted that people are able to efficiently judge the trustworthiness of a face in as little as 100 ms, and this judgment is unlikely to change even if more time is provided. People rely on their intuition in this area.

The following is a list of various behaviors and personality traits that support building trust, as discussed in the literature.

- Ability or competence (Barber, 1983; Luhmann, 1979; Mayer, Davis, and Schoorman, 1995; McKnight, Cummings, and Chervany (1998))
- Benevolence (Luhmann, 1979; Mayer, Davis, and Schoorman, 1995; McKnight, Cummings, and Chervany, 1998)
- Integrity or honesty (Mayer, Davis, and Schoorman, 1995; McKnight, Cummings, and Chervany, 1998)
- Reputation, performance, and appearance (Sztompka,1999)
- Transparent, responsive, caring, sincere, and trustworthy (Bracey, 2002)
- Authentic communication, competence, supporting process, creating boundaries, contact, positive intent, and forgiveness (Bibb and Kourdi, 2004)
- Reliability, consistency, predictability, keeping promises, fairness, loyalty, honesty, discretion, and credibility (Sprenger, 2004)
- Being open, sharing influence, delegating, and managing mutual expectations (Six, 2005)
- Humility, integrity, truth, responsiveness, unblemished fair play, support and encouragement, and team care (Armour, 2007)
- Reliability, openness, competence, and compassion (Mishra and Mishra, 2008)
- Talking straight, showing respect, creating transparency, righting wrongs, showing loyalty, delivering results, getting better, confronting reality, clarifying expectation, practicing accountability, listening first, keeping commitments, and extending trust (Covey, 2009)

It is important to remember that all presented behaviors need to be balanced (i.e., talking straight needs to be balanced by showing respect). Any behavior pushed to the extreme becomes a weakness.

Gefen, Karahanna, and Straub (2003) proposed classifying the antecedents to trust into (1) knowledge-based trust, which focuses on trust building through repeated interactions; (2) cognition-based trust or initial trust, which focuses on trust building through first impressions rather than repeated interactions over a longer period of time; (3) institution-based trust, which focuses on relying on an institution or third party to build trust; and (4) personality-based trust, which refers to individual personalities that influence trust building. Zucker (1986) proposed to identify three mechanisms to build trust. They are (1) process-based trust, which has a similar meaning to knowledge-based trust; (2) characteristic-based trust, which implies that trust is established based on social similarities, such as families, ethnicities, or racial origins; and (3) institution-based trust.

Preparations for the effective process of building trust in the organization should begin from diagnosing the situation in the organization. Sprenger (2009) suggests asking the following questions to help the diagnosis.

- What do managers do to promote a culture of trust in a company?
- What criteria of behavior in organizations allow us to understand the culture of trust?
- What relevant subjects or objects of trust are there in the organization?
- What are the biggest obstacles to building a climate of trust in the company?
- Which of the rules in the organization is the opposite of trust?
- How is the climate of trust created among the leaders?

Researchers have noticed that when employees have trust in the top management, their organizational commitment and organizational identity improve and they work harder and spend more time and energy completing their job functions (Brown and Leigh, 1996; Aryee, Budhwar, and Chen, 2002).

Within organizations, intuition has been posited to help guide a wide range of critical decisions (Dane and Pratt, 2007).

In order for managers to make good decisions, they have to use analysis and intuition as well as trusting themselves. They cannot ignore the available information or their experience. Systematic analysis and intuition are complementary approaches, and how managers use them depends on the situation (Williams, 2012). According to Burke and

Miller (1999) and Swami (2013), many decisions are made unconsciously in the mind. For example, in situations with higher time pressures, increased ambiguities, or a lack of data, creative and quick solutions are needed. As experience and knowledge escalate over time, the decision-making approach becomes more intuitive; therefore, the process turns out to be quicker. In uncertain situations, managers use intuition more often to make decisions (Hyppanen, 2013). According to Hogarth (2010), intuition is shaped by learning. Intuition is considered the outcome of tacit knowledge (Ruggles, 1998), which is learned through experience. Experience can be defined as a set of skills and abilities that an individual accumulates during learning through practice and observation (Malewska, 2013). According to Raskin (1998), we can learn through practice to regain the insights available from intuition. Agor (1989a,b) insisted that intuition is a skill that becomes more prevalent as one moves up the managerial ladder. The use of intuition becomes more common at higher organizational levels than at lower levels. Research suggests that the proportion of executives with intuitive preferences is likely to increase with seniority (Singh, 2009). According to Agor (1989a,b), the following conditions are those in which intuitive decision making seemed to function best.

- When a high level of uncertainty exists
- When little precedent exists
- When variables are less scientifically predictable
- When facts are limited
- When facts do not clearly point the way to go
- When analytical data are of little use
- When several plausible alternative solutions exist to choose from, with good arguments for each
- When the time is limited and there is pressure to come up with the right decision

The role of intuition is enormous. Schreier et al. (2014) confirmed that leaders open to intuition are also open to new experiences. In the literature, it has been suggested that intuition is the first and a necessary stage of creativity (Bastick, 1982; Markeley, 1988). In many creative works, a solution to a problem might come first, and it has to be proved or justified later.

In a data-driven domain, information is disseminated faster and in real-time. This information can be used to predict new problems. However, it is simply human intuition that brings about groundbreaking new ideas. Because of intuition, people are able to identify and predict important trends, opportunities, and threats that lead to innovation.

Parikh (1997, p. 143) argues that creativity is comprised of four elements. They are (1) the capacity for envisioning and understanding intuition, (2) the ability to have a much wider and deeper perception in order to see more than "what meets the eye," (3) the ability to see deeper significance and connections, which may not be obvious, and the ability to break old connections and make new ones, and (4) the skill to convert such connections into concrete applications relevant to the organization and its mission. The author further argues that creativity implies a capacity for vision, intuition, perception, connection, and application.

Using intuition can be beneficial to people. Intuition can build confidence in people's decisions. It can help them make better decisions, develop trust in themselves, have faith in themselves, and strengthen confidence in their own judgments and perceptiveness.

SUMMARY AND DIRECTIONS OF FUTURE RESEARCH

Within the framework of trust and intuition management practices, managers are expected to create the culture that sets work norms and value trust and intuition to produce value-added products and services. Trust allows cooperation between partners who either know each other or are strangers. Trust influences individuals' knowledge and motives. Ambitious goals can only be achieved if people trust each other. This is regardless of whether they know each other or not.

Nowadays, leaders' roles have changed to become more complex and even more critical to the success of organizations. Leaders must handle complexity and ambiguity within organizations, where intuition and trust building become vital. Some leaders can and do deliberately work to improve intuition and build trust in their environment by constantly learning and creating a proper organizational culture. They therefore gain greater confidence in their intuitive powers and their ability to improve the processes within the organization (Patton, 2003). There are limitations to intuitive decision making; however, increasing knowledge of cognitive science and the experiential dimensions with appropriate training can help these limitations. Intuitive decision making will be common, as it is becoming much more widespread in management (Boucouvalas, 1997; Klein, 2003). In conclusion, this review of the literature reveals that the body of knowledge regarding trust and intuition is still limited. The existing literature provides only a fragmented insight into trust and intuition; therefore, further interdisciplinary research in this area is needed.

Several avenues for future research can be suggested that will extend the findings of this research. The literature review has inspired the following research questions for future study:

- How are trust and intuition dependent on the characteristics of individuals and organizational culture?
- How does trust expire in a particular organization?
- How do people rebuild trust in a particular organization?
- How does confidence develop in multicultural environments?

Research related to the cross-cultural perception of trust and intuition as per organizational performance is in its infancy and, therefore, it needs further attention to build theory and practice. By exploring the concept of trust and intuition in different countries, a more global approach to management could be incorporated into strategic planning. Another area for investigation would be to conduct a longitudinal study to determine how trust and intuition are developed (and changed) over time.

REFERENCES

Agor, W. H. (1989a). The intuitive ability of executives: Findings from field research. In W. H. Agor (Ed.), *Intuition in Organizations: Leading and Managing Productively*, (pp. 145–156). Newsbury Park, CA: Sage.

Agor, W. H. (1989b). The logic of intuition: How top executives make important decisions. In W. H. Agor (Ed.), *Intuition in Organizations:. Leading and Managing Productively.* (pp. 157–170), Newsbury Park, CA: Sage Publications.

Andersen, S. M., Glassman, N. S., Chen, S., and Cole, S. W. (1995). Transference in social perception: The role of chronic accessibility in significant-other representations. *Journal of Personality and Social Psychology, 69*(1), 41.

Andersen, S. M., Reznik, I., and Manzella, L. M. (1996). Eliciting facial affect, motivation, and expectancies in transference: Significant-other representations in social relations. *Journal of Personality and Social Psychology, 71*(6), 1108.

Armour, M. (2007). *Leadership and the Power of Trust:. Creating a High-Trust Peak-Performance Organization.* Dallas, TX: LifeThemes Press.

Aryee, S., Budhwar, P. S., and Chen, Z. (2002). Trust as a mediator of the relationship between organizational justice and work outcomes: Test of a social exchange model. *Journal of Organizational Behavior, 23,* 267–285.

Atuahene-Gima, K., and Li, H. (2002). When does trust matter? Antecedents and contingent effects of supervisee trust on performance in selling new products in China and the United States. *Journal of Marketing, 66,* 61–81.

Bachmann, R. (1998). Conclusion: Trust; conceptual aspects of a complex phenomenon. In C. Lane and R. Bachmann (Eds.), *Trust within and between Organizations,* (pp. 298–322). Oxford, UK: Oxford University Press.

Badke-Schaub, P., and Eris, O. (2014). A theoretical approach to intuition in design: Does design methodology need to account for unconscious processes? In A. Chakrabarti and L. T. M. Blessing (Eds.), *An Anthology of Theories and Models of Design:. Philosophy, Approaches and Empirical Explorations,* (pp. 353–370). London: Springer.

Barber, B. (1983). *The Logic and Limits of Trust.* New Brunswick, NJ: Rutgers University Press.

Bartsch, V., Ebers, M., and Maurer, I. (2013). Learning in project-based organizations: The role of project teams' social capital for overcoming barriers to learning. *International Journal of Project Management, 31,* 239–251.

Bastick, T. (1982). *Intuition: How We Think and Act.* New York: Wiley.

Berry, D. S., and Brownlow, S. (1989). Were the physiognomists right? *Personality and Social Psychology Bulletin, 15*(2), 266–279.

Betsch, T. (2008). The nature of intuition and its neglect in research on judgment and decision making. In H. Plessner, C. Betsch, and T. Betsch (Eds.), *Intuition in Judgement and Decision Making.* New York: Taylor and Francis.

Bibb, S., and Kourdi, J. (2004). *Trust Matters For Organizational and Personal Success.* New York: Palgrave Macmillan.

Blau, P. (1964). *Exchange and Power in Social Life.* New York: John Wiley.

Boucouvalas, M. (1997). Intuition: The concept and the experience. In R. Davis-Floyd and P. Arvidson (Eds.), *Intuition: The Inside Story* (pp. 3–18). New York: Routledge.

Bowers, K. S., Regehr, G., Balthazard, C., and Parker, K. (1990). Intuition in the context of discovery. *Cognitive Psychology, 22,* 72–110.

Bracey, H. (2002). *Building Trust:. How to Get It! How to Keep It!* Taylorsville, GA: Hyler Bracey.

Brown, S. P., and Leigh, T. W. (1996). A new look at psychological climate and its relationship to job involvement, effort, and performance. *Journal of Applied Psychology, 81,* 358–368.

Burke, L., and Miller, M. (1999). Taking the mystery out of intuitive decision making. *Academy of Management Executive, 13*(4), 91–98.

Chaiken, S., Liberman, A., and Eagly, A. H. (1989). Heuristic and systematic information processing within and beyond the persuasion context. In J. S. Uleman and J. A. Bargh (Eds.), *Unintended Thought* (pp. 212–252). New York: Guilford.

Colquitt, J. A., Scott, B. A., and LePine, J. A. (2007). Trust, trustworthiness, and trust propensity: A meta-analytic test of their unique relationships with risk taking and job performance. *Journal of Applied Psychology, 92,* 909–927.

Covey, S. M. R. (2009). Building trust: How the best leaders do it. *Leadership Excellence, 22*(2), 22–24.

Dane, E., and Pratt, M. G. (2007). Exploring intuition and its role in managerial decision making. *Source: The Academy of Management Review Academy of Management Review, 32*(1), 33–54.

Das, T. K., and Teng, B. S. (2004). The risk-based view of trust: A conceptual framework. *Journal of Business and Psychology, 19*(1), 85–116.

Dirks, K. T., and Ferrin, D. L. (2002). Trust in leadership: Meta-analytic findings and implications for research and practice. *Journal of Applied Psychology, 87*(4), 611–628.

Dunn, J., and Schweitzer, M. (2005). Feeling and believing: The influence of emotion on trust. *Journal of Personality and Social Psychology, 88*, 736–748.

Encyclopædia Britannica. https://www.britannica.com (accessed 25.03.2018).

Evans, A.M., and Revelle, W. (2008). Survey and behavioral measurements of interpersonal trust. *Journal of Research in Personality, 42*, 1585–1593.

Gefen, D., Karahanna, E., and Straub, D. W. (2003). Trust and TAM in online shopping: An integrated model. *MIS Quarterly, 27*(1), 51–90.

Gibb, J. R. (1978). *Trust: A New View of Personal and Organizational Development.* Los Angeles: Guild of Tutors Press, International College.

Gibbons, D. E. (2004). Friendship and advice networks in the context of changing professional values. *Administrative. Science Quarterly, 49*, 238–259.

Gigerenzer, G. (2007). *Gut Feelings: The Intelligence of the Unconscious.* New York: Viking.

Gilovich, T., Griffin, D., and Kahneman, D. (Eds.). (2002). *Heuristics and Biases: The Psychology of Intuitive Judgment.* New York: Cambridge University Press.

Glaser, J., and Banaji, M. R. (1999). When fair is foul and foul is fair: Reverse priming in automatic evaluation. *Journal of Personality and Social Psychology, 77*(4), 669–687.

Hayashi, A. M. (2001). When to trust your gut. *Harvard Business Review, 79*(2), 59–65.

Hogarth, R. (2010). Intuition: A challenge for psychological research on decision making. *Psychological Inquiry, 21*(4), 338–353.

Hogarth, R. M. (2001). *Educating Iintuition.* Chicago: The University of Chicago Press.

Huang, L., and Murnighan, J. K. (2010). What's in a name? Subliminally activating trusting behavior. *Organizational Behavior and Human Decision Processes, 111*(1), 62–70.

Hyppanen, O. (2013). Decision makers' use of intuition at the front end of innovation. Doctoral dissertation, Department of Industrial Engineering, Aalto University, Finland.

Isaack, T. S. (1978). Intuition: An ignored dimension of management. *Academy of Management Review, October,* 917–921.

Isenberg, D. J. (1984). How senior managers think. *Harvard Business Review, 62*(6), 81–90.

Johnson-George, C., and Swap, W. C. (1982). Measurement of specific interpersonal trust: Construction and validation of a scale to assess trust in a specific other. *Journal of Personality and Social Psychology, 43*, 1306–1317.

Jones, G., and George, J. (1998). The experience and evolution of trust: Implications for cooperation and teamwork. *Academy of Management Review, 23*(3), 531–546.

Jung, G. G. (1924). *Psychological Types,* New York: Harcourt.

Kahneman, D. (2003). A perspective on judgment and choice. *American Psychologist, 58*, 697–720.

Kim, W. C., and Mauborgne, R. A. (1997). Procedural justice, attitudes, and subsidiary top management compliance with multinationals' corporate strategic decisions. *Academy of Management Journal, 36*(3), 502–526.

Klein, G. (1998). *Sources of Power: How People Make Decisions.* Cambridge, MA: MIT Press.

Klein, G. (2003). *Intuition at Work.* New York: Doubleday.

Kramer, R. M. (1999). Trust and distrust in organizations: Emerging perspectives, enduring questions. *Annual Review of Psychology, 50*, 569–598.

Kramer, R. M. and Lewicki, R. J. (2010). Repairing and enhancing trust: Approaches to reducing organizational trust deficits. *Academy of Management Annals, 4*, 245–277

Krumhuber, E., Manstead, A. S. R., Cosker, D., Marshall, D., Rosin, P. L., and Kappas, A. (2007). Facial dynamics as indicators of trustworthiness and cooperative behavior. *Emotion, 7*, 730–735.

Lane, C. (1998). Introduction: Theories and issues in the study of trust. In C. Lane and R. Bachman (Eds.), *Trust within and between Oorganizations:, Conceptual Issues and Empirical Applications* (pp. 1–30). Oxford: Oxford University Press.

Lank, A., and Lank, E. (1995). Legitimizing the gut feel: The role of intuition in business. *Journal of Managerial Psychology, 10*(5), 18–23.

Lewicki, R. J., and Bunker, B. B. (1996). Developing and maintaining trust in work relationships. In R. M. Kramer and T. R. Tyler (Eds.), *Trust in Oorganizations, Ffrontiers of Ttheory and Rresearch* (pp. 114–139). Thousand Oaks, CA: Sage.

Lewicki, R. J., Tomlinson, E. C., and Gillespie, N. (2006). Models of interpersonal trust development: Theoretical approaches, empirical evidence, and future directions. *Journal of Management, 32*, 991–1022.

Lewicki, R. J., Wiethoff, C., and Tomlinson, E. C. (2005). What is the role of trust in organizational justice? In J. Greenberg, and J. A. Colquitt (Eds.), *Handbook of Organizational Justice* (pp. 247–270). Mahwah, NJ: Lawrence Erlbaum.

Lewis, J. D., and Weigert, A. (1985). Trust as a social reality. *Social Forces, 63*(4), 967–985.

Liebowitz, J., Paliszkiewicz, J., and Gołuchowski, J. (Eds). (2018). *Intuition, Trust, and Analytics*. Boca Raton, FL: CRC Press.

Lount, R. B., Jr. (2010). The impact of positive mood on trust in interpersonal and intergroup interactions. *Journal of Personality and Social Psychology, 98*, 420–433.

Luhmann, N. (1979). *Trust and Power*. Chichester, UK: John Wiley.

Malewska, K. (2013). New trends in the decision making process. *Organization and Management, 154*(1), 35–46.

Markley, O. W. (1988). Using depth intuition in creative problem solving and strategic innovation. *Journal of Creative Behaviour, 22*(2), 85–100.

Mayer, R. C., and Gavin, M. B. (2005). Trust in management and performance: Who minds the shop while the employees watch the boss? *Academy of Management Journal, 48*(5), 874–888.

Mayer, R. C., Davis, J. H., and Schoorman, F. D. (1995). An integrative model of organization trust. *Academy of Management Review, 20*(3), 709–734.

McAllister, D. J. (1995). Affect- and cognition-based trust as foundations for interpersonal cooperation in organizations. *Academy of Management Journal, 38*, 24–59.

McEvily, B. (2011). Reorganizing the boundaries of trust: From discrete alternatives to hybrid forms. *Organization Science, 22*, 1266–1276.

McKnight, D. H., Cummings, L., and Chervany, N. L. (1998). Initial trust formation in new organizational relationships. *Academy of Management Review, 23*(3), 473–490.

Mishra, A., and Mishra, K. (2008). *Trust Is Everything:. Become the Leader, Others Will Follow*. Winston-Salem, NC: Aneil Mishra and Karen Mishra.

Molm, L. D., Takahashi, N., and Peterson, G. (2000). Risk and trust in social exchange: An experimental test of a classical proposition. *American Journal of Sociology, 105*, 1396–1427.

Myers, D. G. (2002). *Intuition: Its Powers and Perils*. New Haven, CT: Yale University Press.

Nooteboom, B. (2002). *Trust: Forms, Foundations, Functions, Failures and Figures.* Cheltenham, UK: Edward Elgar

Olekalns, M., and Smith, P. L. (2005). Moments in time: Metacognition, trust, and outcomes in dyadic negotiations. *Personality and Social Psychology Bulletin, 31,* 1696–1707.

Paliszkiewicz, J., and Koohang, A. (2016). *Social Media and Trust: A Multinational Study of University Students.* Santa Rosa, CA: Informing Science Press.

Paliszkiewicz, J., Koohang, A., Gołuchowski, J., and Horn Nord, J. (2014). Management trust, organizational trust, and organizational performance: Advancing and measuring a theoretical model. *Management and Production Engineering Review,* (1), 32–41.

Parikh, J. (1997). *Managing Your Self:. Management by Detached Involvement.* Oxford, UK: Blackwell.

Patton, J. R. (2003). Intuition in decisions. *Management Decision, 41*(10), 989–996.

Plessner, T., Betsch, C., and Betsch, T. (Eds.). (2008). *Intuition Judgment and Decision Making.* Mahwah, NJ: Erlbaum.

Policastro, E. (1999). Intuition. In M. A. Runco and S. R. Pritzker (Eds.), *Encyclopedia of Ccreativity,* vol. 2 (pp. 89–93). San Diego: Academic Press.

Raidl, M. H., and Lubart, T. I. (2000–2001). An empirical study of intuition and creativity. *Imagination, Cognition and Personality, 20,* 217–230.

Raskin, P. (1998). Decision-making by intuition, Part 1: Why you should trust your intuition. *Chemical Engineering, November. 21,* 100–102.

Righetti, F., and Finkenauer, C. (2011). If you are able to control yourself, I will trust you: The role of perceived self-control in interpersonal trust. *Journal of Personality and Social Psychology, 100*(5), 874–886.

Rotter, J. B. (1967). A new scale for the measurement of interpersonal trust. *Journal of Personality, 35,* 651–665.

Rotter, J. B. (1971). Generalized expectancies for interpersonal trust. *American Psychologist, 26,* 443–452.

Rousseau, D. M., Sitkin, S. B., Burt, R. S., and Camerer, C. (1998). Not so different after all: A cross-discipline view of trust. *Academy of Management Review, 23,* 393–404.

Ruggles, R. (1998). The state of the notion: Knowledge management in practice. *California Management Review, 40,* 80–91.

Sadler-Smith, E. (2008). *Inside intuition.* Abingdon, UK: Routledge.

Scharlemann, J. P., Eckel, C. C., Kacelnik, A., and Wilson, R. K. (2001). The value of a smile: Game theory with a human face. *Journal of Economic Psychology, 22*(5), 617–640.

Schreier, C., Schubert, A., Weber, J., and Farrar, F. (2014). An investigation of the character traits of decision-makers open to intuition as a tool. *GSTF Journal on Business Review, 3*(4), 63–69.

Shapiro, S., and Spence, M. (1997). Managerial intuition: A conceptual and operational framework. *Business Horizons, 40*(1), 63–68.

Shirley, D. A., and Langan-Fox, J. (1996). Intuition: A review of the literature. *Psychological Reports, 79,* 563–584.

Simpson, J. A. (2007). Psychological foundations of trust. *Current Directions in Psychological Science, 16,* 264–268.

Singh, A. (2009). Leadership grid between concern for people and intuition. *Leadership & and Management in Engineering, 9*(2), 71–82.

Six, F. (2005). *The Trouble with Trust: The Dynamics of Interpersonal Trust Building .* Bodmin, UK: MPG Books.

Sprenger, R. K. (2004). *Trust:. The Best Way to Manage.* London: Cyan/Campus.

Sprenger, R. K. (2009). *Zaufanie* #1 [Trust #1]. Warsaw, Poland: MT Biznes.

Swami, S. (2013). Executive functions and decision making: A managerial review. *IIMB Management Review, 25*(4), 203–212.

Sztompka, P. (1999). *Trust: A Sociological Theory.* Cambridge, UK: Cambridge University Press.

Tyler, T. R., and Kramer, R. M. (1996). Wither trust? In R. M. Kramer and T. R. Tyler (Eds.), *Trust in Organizations: Frontiers of Theory and Research* (pp. 1–15). Thousand Oaks, CA: Sage.

Van't Wout, M., and Sanfey, A. G. (2008). Friend or foe: The effect of implicit trustworthiness judgments in social decision-making. *Cognition, 108*(3), 796–803.

Westcott, M. R., and Ranzoni, J. H. (1963). Correlates of intuitive thinking. *Psychological Reports, 12,* 595–613.

Wild, K. W. (1938). *Intuition.* Cambridge, UK: Cambridge University Press.

Williams, K. C. (2012). Business intuition: The mortar among the bricks of analysis. *Journal of Management Policy and Practice, 13*(5), 48–65.

Willis, J., and Todorov, A. (2006). First impressions making up your mind after a 100-ms exposure to a face. *Psychological Science, 17*(7), 592–598.

Wong, Y. T., Wong, C. S., and Ngo, H. Y. (2002). Loyalty to supervisor and trust in supervisor of workers in Chinese joint ventures: A test of two competing models. *International Journal of Human Resource Management, 13*(6), 883–900.

Worchel, P. (1979). Trust and distrust. In W. G. Austin and S. Worchel (Eds.), *The Social Psychology of Intergroup Relations* (pp. 174–187). Belmont, CA: Wadsworth.

Wrightsman, L. S. (1966). Personality and attitudinal correlates of trusting and trustworthy behaviors in a two-person game. *Journal of Personality and Social Psychology, 4,* 328–332.

Yamagishi, T., Cook, K. S., and Watabe, M. (1998). Uncertainty, trust, and commitment formation in the United States and Japan. *American Journal of Sociology, 104,* 165–194.

Zand, D. (1972). Trust and managerial problem solving. *Administrative Science Quarterly, 17,* 229–239

Zucker, L. G. (1986). Production of trust: Institutional sources of economic structure. In B. M. Staw and Cummings L. L. (Eds.), *Research in Organizational Behavior,* (pp. 53–111). Greenwich, CT: JAI Press.

Zucker, L. G., Darby, M. R., Brewer, M. B., and Peng, Y. (1996). Collaboration structure and information dilemmas in biotechnology: Organizational boundaries as trust production. In R. M. Kramer and T. R. Tyler (Eds.), *Trust in Organizations, Frontiers of Theory and Research,* (pp. 90–113). Thousand Oaks, CA: Sage.

Part Two

Examples of Executive Decision Making and Intuition

5

Filtering Intuition through the Lens of Rational Analysis: Decision-Making Lessons Learned in a Competitive Higher Education Market

Written by: Susan C. Aldridge

CONTENTS

As in any competitive industry, university leaders must use such multi-dimensional skills as critical thinking, change management, and strategic innovation to effectively meet the increasingly diverse and ever-evolving needs of the students and boards they serve, as well as the people they employ. In fact, we work in a highly complex system, where there is often no one best solution, given the "wicked" challenges that come with limited resources and conflicting priorities, competing ideologies and a complicated regulatory environment (Aldridge and Harvatt, 2014). Consequently, intuition is a valuable asset and one that I have come to rely on in a professional field where decisive action and creative problem-solving are essential to success.

Although rational analysis certainly helps inform my decision-making process, so does experience and expertise, inner vision, self-awareness, and collaborative spirit—all of which, when combined, have enhanced by ability to turn gut feelings into big ideas and big ideas into innovative solutions. Like many executives, I routinely employ data to help frame major decisions that involve significant investments of time and money, using it to pinpoint the challenge, structure a narrative for defining it, and confirm that the selected approach is effective in meeting it. On the other hand, intuition provides those all-important insightful moments that help visualize patterns and formulate concepts that result in good, working theories to use for taking strategic action in a competitive market.

To be sure, intuition is not simply a random or mystical process but rather a well-developed perceptual skill grounded in relevant experience and expertise, which can be cultivated and refined over time (Matzler, Ballom, and Mooradlan, 2007; Krulak, 1999). Research has also shown that when honed through the right experience and expertise, intuitive prowess can be just as effective as analytical skills in the executive decision-making process, particularly in complex, ever-changing environments (Dane, Rockman, and Pratt, 2012; Mousavi, 2014). So, by accessing my intuition and supporting it with available data, I have learned to trust and exploit both for making executive decisions in an industry I have spent the bulk of my career: adult and distance higher education.

That implicit knowledge served me well in 2006, after assuming the presidency of University of Maryland University College (UMUC), one of 13 degree-granting institutions within the state's public university system, created to fulfill a distinctive mission. Unlike its sister schools, UMUC welcomed all qualified applicants and catered primarily to adult working professionals, nearly half of whom were military students. To serve their unique needs, the university had built a massive online education enterprise, in addition to offering in-person academic options across three divisions in the United States, Europe, and Asia through satellite sites in the contiguous domestic states and on military bases around the world.

Having already successfully led a similar (although smaller) global campus at Troy University in Alabama, I was well versed in the demographics, motivations, and behaviors of the adult students UMUC enrolled. These working professionals were part of a rapidly growing population of so-called non-traditional students: men and women from all walks of life and many cultural traditions who returned to college for any number of reasons—to earn a degree or update a credential, to get ahead in their profession or change careers altogether, to learn something new or pursue an ongoing interest.

Moreover, these students were often balancing the demands of school with the responsibilities of work and family life, which for military service members included frequent transfers and deployments. Thus, they were in search of a seamless, flexible, relevant, and student-centered academic experience—full time and part time, online and in person, with industry-focused programs and curricula, as well as liberal transfer credit options and easily accessible support services.

In addition to understanding the student population, I believed that UMUC had built its reputation in the adult-serving higher education market around both an academic portfolio and an institutional profile that checked all of these boxes. But I had also learned from experience that the higher education landscape was changing rapidly, and as a result, I would need to take a hard look at UMUC's present state to strategically plan for its future. That said, I conducted plenty of research before accepting the position as president, reviewing and analyzing all of the available data.

Yet while that rational assessment helped inform my decision, my intuition led me to the opportunity to serve the adult students who needed an educational champion. I simply *knew* in my gut that it was the right move, based on my experience and expertise, as well as my entrepreneurial skills and collaborative nature. In fact, it didn't take long to envision myself in this new role, using those attributes to help my new colleagues critically and strategically assess the environment and use design thinking skills in an ever more complex and competitive market. This same intuitive decision-making capacity also made it possible to successfully power through and beyond the reality I faced once I actually stepped into that role—a reality the data had failed to reveal.

A MULTITUDE OF CHALLENGES IN A SEA OF OPPORTUNITY

As UMUC's new president, I learned in my second week that the university was facing some serious challenges that were having a major impact on revenue growth, 90% of which was tuition dependent. After analyzing yet more data and eliciting information from the senior leadership team, I quickly identified a number of significant issues that were driving these challenges.

To begin with, by providing both unprecedented access and far greater economies of scale, the university's pioneering move into online education had created a "perfect storm" for burgeoning enrollments throughout the nineties and into the first few years of the twenty-first century. This rapid growth placed tremendous stress on an institutional infrastructure that was not yet sophisticated enough to support it. As a result, the university struggled to keep up with rapidly escalating enrollment demand at a time when the for-profit higher education industry was becoming a tsunami in the adult-serving market. In fact, the previous fall semester's enrollments were some 5000 students less than anticipated. On top of that, UMUC's management structure was largely decentralized, with each of its three divisions and many of its functional departments operating independently of one another, thereby making it even harder for students to move seamlessly through the global university.

Likewise, the admissions process was cumbersome and paper heavy, particularly when it came to processing financial aid, degree audits, and credit transfer requests. Thus, by the end of 2005, students were waiting nearly 3 months to enroll; and transfer students anywhere from 4 to 8 months to obtain an official credit evaluation. Not surprisingly, this burdensome process was causing an ever-growing number of prospective students to simply give up, or worse yet, enroll in one of the faster-moving for-profit universities.

The military was also downsizing, which portended declining military student enrollments and base closures, while the state had cut millions from its higher education budget. Even more problematic, UMUC had neglected to continue investing in high-impact signature programs that would attract large numbers of students. As a public university that built its reputation around providing working adults with marketable credentials, it was losing momentum to its competitors when it came to launching career-relevant programs that would further enhance its brand and boost its enrollments.

Consequently, when I assumed the presidency in February 2006, UMUC was facing a variety of significant challenges that threatened its long-term viability, the most urgent of which, by far, was the need to enroll 9000 new students for fall semester, to meet the state's growth targets. What's more, according to its CFO, the university's financial situation was dismal, which meant urgent budget cuts to garner the requisite funds for recruiting students. The university would therefore need to significantly step up enrollment growth going forward, to not only stabilize revenues but also expand them considerably for future investment.

As a longtime administrator in adult and distance higher education, I had navigated most of these challenges in one form or another, by charting a new course through the sea of opportunities. Consequently, the team would have to question the status quo and take acceptable risks around continuous innovation that would pave the way for UMUC to grow and thrive in the face of rapid and uncertain change. So, based on both the available data and what I intuitively knew about the students we served, I concluded that once we met the immediate enrollment targets (which was going to be a herculean feat), we would tackle two priority initiatives, designed to re-sharpen the university's competitive edge.

The first priority was streamlining the admissions process, to greatly improve its efficiency and dramatically shorten its time frame. Once this initiative was successfully underway, we would then identify and develop a few groundbreaking signature programs that played directly into UMUC's brand attributes and reputation, while also attracting thousands of new enrollments in a relatively short time.

BREAKING THROUGH THE ADMISSIONS LOGJAM

I had barely settled into my office before taking on the arduous task of meeting the enrollment target of 9000 new students. Because there was no one department or office responsible for accomplishing this objective, a university-wide enrollment task force was deployed, comprising faculty and staff representatives from across the university and at every level within its hierarchy. Although the immediate objective was to resolve the crisis at hand, the ultimate purpose was to come up with a far better enrollment management process.

As the designated leader of this task force, I empowered its members to push the envelope, through open and honest collaboration across divisions, departments, and disciplines, with the goal being to evaluate every step in the admissions process. To fuel this effort, we established a weekly meeting schedule, along with two simple ground rules: all ideas were on the table, and all voices would be heard. Each member of the team also agreed to assume individual responsibility for his/her decisions and actions, while at the same time working collectively to achieve them.

In the weeks that followed, these dedicated university employees mapped the admissions process to identify roadblocks, determined to succeed in spite of daunting odds. And within just 6 months, we not only met our fall enrollment target, we exceeded it, bringing in 10,000 new students— more than were enrolled on all but a few campuses in Maryland. With the enrollment emergency behind us, we were able to measure its impact, an exercise that resulted in more than a few moments of critical feedback around how we could improve the student experience.

The first big hurdle was to overhaul enrollment management in line with UMUC's emerging growth and development strategy. We continued mapping the business process with an eye toward enhancing the student experience, identifying viable practices and eliminating major pain points—a long-time habit that helped fuel my intuition around where to invest and divest, as well as what to retain or jettison. I also encouraged the team to share their own experiences, big ideas, and relevant data around how this process could be improved. As a result, we collectively accomplished our initial objective by

- Ensuring a university-wide, shared ownership in student recruitment, experience, and retention
- Creating a well-trained and customer service-focused team of enrollment management professionals, with a clearly defined career path
- Assigning enrollment advisors to specific student subpopulations, such as military service members and first-generation college students
- Developing an enrollment call center, using a highly scalable model for customized student outreach, which enabled 24/7/365 student interaction—online, by phone, or in person
- Hosting weekend and evening open houses for prospective students, with the help of faculty and alumni volunteers
- Establishing service standards and metrics that were constantly measured and visible to the president

In addition to restructuring the enrollment services team and reinventing its approach, we explored how the university might improve financial aid processing, which was not only paper heavy but also labor intensive, given that financial aid packaging and approval was done primarily by hand. Moreover, each division operated separately within the context of its own set of business practices. For working adults juggling other major financial responsibilities, enrollment often depends on offsetting out-of-pocket tuition payments with a variety of assistance options, from student loans and Pell grants to military benefits and employer reimbursements. Consequently, the data we gathered from the staff only confirmed what

I had surmised: UMUC's cumbersome financial aid process, without defined quality metrics, was further slowing our enrollment growth.

Having dealt with a similar issue at my former university in Alabama, I had learned that automating major facets of the process would lead to greater efficiencies and expanded enrollments. Likewise, service standards and weekly data monitoring, along with staff restructuring and training, were essential. In a matter of months, UMUC had metrics in place to track service standard and time and cost efficiencies within the bigger picture of our newly streamlined admissions process. In addition, the university transitioned faculty recruitment to human resources (HR) on a year-round schedule and instituted new practices for adding course sections, as needed. Both of these areas were vital for scalability.

Once these improvements and others were underway, it was clear that by framing the challenge with appropriate data, while trusting intuition grounded in professional experience with the adult student population, the team was inspired to creatively come up with a far better system for serving UMUC's students and simultaneously meeting its growth and development strategy. In fact, together we had reduced the average number of days from initial inquiry to admission from 81 to 14, within only 6 months. But as dramatic as that improvement was, we were still lagging behind in the transfer credit and degree audit process.

Once again, automation would be the key to our success. Of course, this was a far more complex process than financial aid, which would mean an exponentially greater investment of dollars and staff resources. On the other hand, it was an investment with a potentially enormous return. Over the years, I had heard far too many horror stories from students who, in transferring from one institution to another, lost out on scholarships, added needlessly to growing student loan debt, and/or had to postpone highly anticipated graduation dates, simply because it had taken far too long to evaluate and confirm course transfer credits. And like its institutional peers, UMUC was still operating in *snail time* when it came to transfer credit evaluation and degree auditing across its three divisions. Therefore, a state-of-the-art system would enable the university to move from *snail time* to *real time*, a strategic differentiator for the university that, by setting it apart in a highly competitive market, would go a long way in achieving its growth and development objectives.

As with its financial aid process, UMUC relied on paper-based file systems that were costly and cumbersome to use and store, as well as extremely vulnerable to natural and manmade disasters. But even more important, there was no easy or even consistent way of sharing transfer information among schools or even across university divisions—a major problem for adult students, who often arrived with multiple transcripts from other institutions they had previously attended. In fact, military students submitted, on average, five transcripts; and as they moved between assignments, their transcripts were being mailed from one UMUC division to the next. Consequently, it typically took weeks and sometimes months for prospective students to obtain much-needed information around the courses required to complete a degree and graduate on time. Likewise, UMUC's antiquated system cost students who had enrolled precious time and money taking courses they didn't need; and in far too many instances, the university eventually lost them altogether.

As the new leader of an institution with enormous enrollment and financial challenges, my focus on automating student services was unorthodox. Yet in my mind, any major barrier to admissions was unacceptable. With this challenge on the table, the university became determined to design and implement a paperless system that would quickly and efficiently process these transfer credits. After tasking a "dream team" of key UMUC staff with turning this idea into an innovative solution, Hershey Systems was selected as the development partner, in light of the company's reputation as the leading provider of student-centered software solutions.

In putting our students and their needs front and center, we envisioned a system that would ultimately generate customized academic portfolios, thus providing both prospective and current students with a user-friendly online tool for plotting their academic trajectory and protecting their vital information. This tool would allow them to obtain unofficial course credit evaluations in real time, much as they would an ATM bank balance. Yet while we needed a system that made effective use of technology, we also wanted one that could accommodate students who needed the "human touch" of a faculty advisor, from time to time.

Finally, we were determined to create a mechanism that would leave no institution behind, given that students requested credit transfers from a multitude of community colleges and military training schools, other 4-year universities, and prior real-world learning experiences. Consequently, to gather and process information from anywhere in the world, we would need an automated system that was accessible and scalable, reliable and secure.

A year later, we unveiled a system that quickly became a strategic differentiator for the university. In fact, as a result of our collective vision and innovative mindset, our intuitive knowing and rational analysis, it proved to be a winning solution. By drastically reducing the time it took to process degree audits and credit transfer requests—from as long as 8 months to, on average, 48 hours—it saved transfer students untold dollars and countless hours taking classes they might not otherwise need. The faculty and staff embraced the new system as well, given that it greatly enhanced their productivity and reduced the tedium of juggling paper. It also furnished enrollment counselors and academic advisors with ready access to information they needed for helping prospective and current students create and maintain a realistic degree completion plan.

Likewise, it provided UMUC with a variety of operational benefits. Within a relatively short time, the university tripled the number of official audits it conducted each month, reduced the cost per audit, and did away with both the risk and the expense of paper records storage. This system also played a significant role in an aggressive enrollment growth and student retention strategy. By producing an unofficial credit evaluation in real time, prospective students could take a quick peek into their academic future with UMUC, which, in turn, made it easier to convert initial inquiries into confirmed enrollments. What's more, it enabled students to move through their degrees, by giving them a handy tool for tracking their trajectory all the way to graduation. But even more important, by increasing the university's efficiencies of scale overall, everyone felt empowered to serve students far more effectively.

LAUNCHING THE NEXT BIG IDEA AT JUST THE RIGHT TIME

Having launched a much-improved admissions process, the university turned its full attention to the second priority initiative I knew we must vigorously pursue: a few groundbreaking signature academic programs. Selecting the right signature program would require a keen sense of timing and urgency, given that *first-to-market* advantage can never be underestimated. In addition, having extensive experience and expertise around the competition, as well as the unique attributes of the students and the institution, was critical to success.

Unlike traditional students, adults return to college primarily for professional advancement, and in tight job markets that often means changing careers altogether. Likewise, universities are always in search of opportunities to differentiate themselves from their institutional peers, and signature programs offer a tremendous return on investment, when chosen carefully. To be sure, by providing students with a unique opportunity to move into or ahead in a high-demand professional field, these programs often yield significant enrollments in a relatively short time, especially when they align well with a university's brand.

Thus, it was vitally important to identify signature offerings that were a good fit for UMUC's distinctive combination of brand assets and attributes, which included

- A unique mission to serve adult working professionals
- A 60-year legacy serving U.S. military service members, their families, and veterans around the world
- The first university in the United States to offer online courses and one of the largest and fastest-growing online education enterprises in the world, with an ability to quickly develop, launch, and scale a virtual program to add thousands of new enrollments
- A reputable public university within a short distance to the nation's capital
- A solid cadre of adjunct and full-time scholar-practitioner faculty, with extensive experience in their professional fields
- An extensive network of governmental, institutional, and corporate partners
- Among the largest portfolios of computer science and information technology (IT) degree programs on the country's eastern seaboard, along with solid offerings in science and business administration
- One of fewer than 150 universities in the United States designated by the National Security Agency as a Center of Academic Excellence in Information Assurance Education

With a profile like that, there were more than a few signature program candidates with excellent potential, particularly in STEM and business disciplines that were a good fit for the university's student population, including its large number of military students. Given its adult-serving mission and student demographics, UMUC had been among the first major universities to invest heavily in computer science and IT programs as the demand for qualified IT professionals began to escalate in the 1980s. It had also distinguished itself in other science fields, as well as in the management arena. Thus, it had a relatively broad spectrum of academic areas from which to choose a few good options.

To launch its signature program initiative, UMUC reengineered its master of arts in teaching (MAT) program to provide career changers (including retiring military service members) with a novel alternative pathway to teacher certification, at a time when there were jobs to be filled in Maryland. Aimed at preparing high school teachers in critical STEM shortage areas, this program enrolled students who earned their certificates by completing six online courses, a nine-credit teaching internship, a professional seminar, and an onsite residency. Then, in 2008, the university became one of the first in the world to pioneer the professional science master's (PSM) degree, which could be completed entirely online. Touted by many at the time as the "new MBA," this degree combined graduate studies in science or math with business management courses, in such high-demand career areas as biotechnology, environmental management, and information assurance. Consequently, it enrolled more than 500 students within a few short months.

Yet while these programs were attracting a fair number of enrollments, UMUC needed a bolder idea to secure thousands more students in the short term, and my intuition told me that cybersecurity was a timely opportunity, based on what I was hearing from our government contacts. While the country desperately needed some 30,000 cybersecurity professionals with advanced skills, the current talent pool was severely limited to around 1000 such individuals, given that there were few if any institutions focused on cybersecurity per se as an emerging academic discipline. At the time, the university lacked sophisticated market research capabilities to corroborate my intuition.

Given UMUC's brand, its students and its online course scalability, I began to imagine cyber as the next big academic frontier for rapid enrollment growth. To be sure, it would not only play into the university's strengths, it would also lend itself well to both undergraduate and graduate degrees and certificates of signature status. What's more, with robust course development, these options could be customized for multiple industries beyond the government's national security sector, including healthcare, transportation, finance, energy, education, and criminal justice. And given the university's proven ability to ramp up new programs in a relatively short time, I was certain that it could lead well ahead of the competition. But before acting on my intuition, solid research was essential.

After gathering data and talking to both board members and experts in the cybersecurity field, I zeroed in on some of the unique challenges that both government agencies and corporations faced in building a solid pool of well-trained cyber-warriors. Although there were a few respected universities providing courses and degree specializations in the information security field, none offered a continuum of full-scale undergraduate and graduate degree programs in cybersecurity. Equally problematic, training furnished through industry associations was far from scalable, as it was in person during the day, rather than online at any time, thereby taking employees away from their work. Moreover, given the extremely limited talent pool, companies were spending, on average, $40,000 to recruit a qualified cybersecurity professional, which had to be accomplished through job fairs, face-to-face interviews, and print ads. Likewise, it was difficult for employers to quickly obtain security clearances for the experts they did hire, unless they arrived on the job with one in hand.

After analyzing extensive data, the defining attributes of both marketable cybersecurity programs and prospective students became clear. These programs would have to be offered predominantly (if not fully) online, with a "bleeding edge" curriculum mapped to both government and corporate requirements, as well as highly interactive courses built around real-world case studies. A secure virtual lab was essential for hands-on practice from any global location, along with a first-rate faculty comprising well-respected scholar-practitioners in the field.

Over the years, I had learned that the best way to design career-relevant programs was to convene a "think tank" of industry experts, who by working closely with the university's academic team could provide exceedingly beneficial input and feedback around everything from curriculum to marketing. Think tank members were also invaluable for opening doors to private- and public-sector partnerships that could yield a sustainable pipeline of new students for years to come.

Working through a network of professional connections, we created a powerful think tank comprising the country's top authorities in cybersecurity, including several who had served as advisors to numerous U.S. presidents. Think tank members were tasked with reviewing the proposed curriculum to ensure that it closely reflected cybersecurity industry standards and expectations, while incorporating courses that demanded the utmost intellectual rigor, both of which were critical to the university's success. With this big idea now firmly in mind and a solid group of experts with whom to fully explore and develop it, it was essential to obtain "buy-in" from UMUC managers and faculty.

After extensive discussions around the data we were using to frame the rationale for these programs, I invited senior administrators, faculty, and staff to attend a daylong meeting with think tank members. After listening to the experts speak both passionately and intelligently about UMUC's tremendous potential for successfully launching cybersecurity programs, they quickly galvanized behind the provost, who would take the lead in developing the curriculum and obtaining the necessary approvals from the state.

Working at full speed, this collaborative alliance of industry and academic experts suggested two master's degree programs in cybersecurity and cybersecurity policy, a bachelor's degree program in cybersecurity, and at least three graduate certificate programs—all of which would be delivered predominantly online to maximize student outreach. With these offerings in mind, they joined forces to draft the curriculum and ensure that it met industry standards around much-needed advanced skills. The provost then contacted HR departments in a carefully selected group of major corporations, asking them to review the curriculum and suggest modifications.

After obtaining state approval, the university hired subject matter experts from across the country to work with instructional designers and faculty around developing media rich and academically rigorous courses of the highest quality, with first-rate graphics and simulations that provided an active and authentic learning experience. We built a unique, remote-access Cyber Virtual Lab, recruited an exceptionally talented and credentialed group of faculty members from some of the best universities in the United States, and designated a cyber-specific research librarian to serve online students. Likewise, more than $1 million in merit-based cybersecurity scholarship funds were raised and we began courting partnerships among government agencies and private companies that would most likely send their employees to UMUC with tuition reimbursement in hand.

A year into program development, the governor of Maryland helped move our student recruitment efforts a few steps forward when he unexpectedly announced his plan to make Maryland the nation's cybersecurity center of excellence—a move that could bring as many as 28,000 jobs right into UMUC's backyard. Indeed, having a ready-made and easily accessible portfolio of high-quality cybersecurity programs, the university rapidly became a valuable state asset for recruiting companies to relocate or expand. To make things even easier for these companies, UMUC used its expertise in the online space to create virtual career fairs open to its cybersecurity students, which greatly reduced both the time and the cost for them to recruit qualified professionals with advanced skills. What's more, when corporate recruiters realized that UMUC's cybersecurity students were indeed workforce ready, they forged solid partnerships with the university that became essential not only for recruiting new professionals but also for upgrading skills among their current employees.

During the program's second year, the university further solidified its academic leadership in the rapidly emerging cybersecurity discipline by hosting a daylong "meeting of minds" that focused on critical public policy initiatives under consideration both in the United States and Canada. By generating interest—and attendance—among government officials and corporate leaders in both countries, it provided an additional opportunity to cultivate significant enrollments and program investment funding. The university also entered the cybersecurity competition arena with its award-winning Cyber Padawans, a team of students, faculty, and alumni who began acquiring numerous honors in some of the world's more prestigious competitions—another way to attract enrollments.

Of course, the proof that my intuition had paid off rested in the data constantly analyzed to measure our success, and the outcomes far exceeded even my expectations when it came to breaking records in new student enrollment. Within 18 months of unveiling this initiative, UMUC had already received more than 5300 applications and enrolled some 3000 truly outstanding students, enrollments that represented 4% of the university's global revenue. Some 90% of the graduate students had at least 5 years' professional experience in IT, information assurance, computer security, or cybersecurity across all sectors of the economy. Equally impressive, nearly half of them had earned an industry or professional certification in their field, and 61% already held a security clearance.

This data also set the stage for UMUC to develop a third graduate degree in digital forensics and cyber-investigation, along with a Cyber 101 course for all incoming students that served as a basic introduction to cybersecurity awareness and protection—important skills in today's digital technology–driven world. In addition, we began mapping specific cybersecurity concentrations across a variety of relevant industries. Consequently, our fast and effective foray into cybersecurity energized the university community in many new directions. Not only was UMUC collaborating with some of the country's most distinguished cybersecurity experts, it had also positioned itself as an academic leader in this high-demand professional field.

With these priority initiatives in place—a much streamlined admissions system and a portfolio of new and truly powerful signature programs—the university was well on its way to meeting its strategic growth and development goals. But neither of them would have been possible without marshaling the discipline to cut budgets and reallocate funds to support them, which required making extremely difficult decisions along the way.

FINAL THOUGHTS

In retrospect, I can say that what began as solid hunches, grounded in both intuitive knowing and good data, inspired both the fast thinking and creative problem-solving UMUC needed to lead in an ever more complex and competitive higher education market. Yet while I believed we were on the right path, it was important to continue filtering intuition through the lens of rational analysis, using both to envision the outcome we wanted to achieve and develop the key performance indicators (KPIs) needed to track our progress. In choosing the right KPIs, it was important to ensure that they not only aligned with our goals but also were accurate, attainable, and actionable. In other words, they were based on legitimate data, provided a practical process and context for reliable measurement, and enabled us to achieve the desired result.

What's more, while I rely on both intuitive and analytical skills in moving forward with a big idea, I am also constantly alert to the vital role they play in signaling when it is time to readjust a course of action or even abandon it altogether. Indeed, there have been more than a few occasions over the years when my instinct warned, and the data confirmed, that what seemed to be an effective plan at the time was going awry for one reason or another. So, as the university moved ahead with its growth initiatives, I periodically took a step back to ask the right questions and gather any additional information needed to challenge assumptions, evaluate decisions, and, if warranted, change direction.

At the same time, university team members were rewarded for employing their intuitive and analytical skills, to improve both student services and academic program offerings. We cultivated new managers—imbued with experience, expertise, and inner vision; self-awareness, analytical prowess, and a collaborative spirit—and empowered them to use these attributes wisely. Having the freedom to act on their unique experience and novel patterns of thinking, while utilizing relevant data to frame the issue and monitor the outcome, they seeded innovative solutions in their own functional areas, which, in turn, helped round out the "big picture" perspective needed to keep the institution moving forward. Thus, by collectively flexing our intuitive and analytical muscles, we not only powered through UMUC's immediate challenges, we set its course for the future. In just a few short years, UMUC had added millions of dollars to its investment fund and doubled its annual revenue. It had also increased its enrollments to around 97,000 students, thus becoming the nation's largest public university in a fiercely competitive adult-serving market.

REFERENCES

Aldridge, S., and K. Harvatt (2014), *Wired for Success*, p. 20. Washington, D.C.: American Association of State Colleges and Universities.

Dane, E., K. Rockman, and M. Pratt (2012), "When should I trust my gut? Linking domain expertise to intuitive decision-making effectiveness." *Organizational Behavior and Human Decision Processes*, p. 187–194, 119 (2).

Krulak, C. (1999), "Cultivating intuitive decision making," *Marine Corps Gazette*, May.

Matzler, K., F. Bailom, and T. Mooradian (2007), "Intuition decision making," *MIT Sloan Management Review*, p. 13–15 Fall.

Mousavi, S. (2014), "Risk, uncertainty, and heuristics," *Journal of Business Research*, p. 1671–1678, 67 (8).

6

Trusting Intuition

Written by: R.T. Good

CONTENTS

Some assume that relying on intuition for executive decision making is to be devoid of other information. When characterized with a positive spin, we say that one has a "gut feeling." When not so positive, we conclude that one is "shooting from the hip." Interesting that both physical references to the body are not so far apart, suggesting that they come from a similar place of origin.

In my experience, I have come to believe that *effective* reliance on intuition contains two important components: (1) that it does not happen in a vacuum, and (2) that precision in using intuition is a skill honed by practice that cultivates attention, openness, and trust born of genuineness. Moreover, my experience has been that, regardless of outcome, as a result of relying on intuition for decision making, I live more comfortably with the outcome by virtue of the process used in rendering the decision.

In this chapter, I endeavor to share some of my personal perspectives and experiences in trusting my intuition for executive decision making. In doing so, I will address what I see as the counterbalance of thinking on the process of making executive decisions, how the two thought streams can act in conflict or concert, and what I have found to be an approach for making choices that create synergy between the two and which has been effective for me. In the end, I hope to share a perspective from which others might draw insights that enliven their own reliance on the value of and practices for trusting intuition.

THE DICHOTOMOUS QUESTION

Facts or feeling: on which can we depend for making decisions? The debate is often won on the frequently accepted tenet "Facts don't lie." While there is increased attention being deservedly paid to the value of decision making improved by data analytics that results from increased computing power, we may simultaneously be starting to trend toward data-driven decision making absent of the larger contextual picture. Such seems to be the case with the recent precipitous and record-breaking drop of the U.S. financial markets driven by algorithmic trade determinants. While the facts didn't lie, the decisions generated were determined to be extreme and faulty (Harwell, 2018). No wonder there is pause by some to surrender to complete automation the control of our transportation, our livelihoods, and other tools of existence that were formerly supervised by human interpretation.

Concurrently, at the other extreme, U.S. politics seems to have devolved into a period of emotional posturing that primarily drives policy decisions—or the lack thereof. No longer are the institutions of investigation relied on to reveal the truth from which policy is formulated; rather, all truths are judged relative to one's worldview, biases, and agendas. To the extent that this is perpetuated by the electorate or the elected, the right or the left, I will leave for others to debate. While debate is healthy, the paralysis of progress driven primarily by entrenched feelings is problematic. It may be surprising to learn of a recent initiative wherein members of Congress are asked to take a pledge to "join with colleagues on both sides of the aisle on at least one piece of major legislation each year," and to "meet with someone from the opposing party one-on-one at least once a month" (Mullen and Ackerman, 2018). While this is reported as a positive development, the meager commitment to which congressional leaders are asked to pledge is problematic, falling far short of what I believe is expected by the populace of our elected officials and indicative that decision making is essentially non-existent. Consequently, at either extreme, there is a danger of a lack of effectiveness when decisions are rendered solely by either facts or feelings.

THE MIDDLE WAY

But what of intuition? I contend that intuition is the careful integration of both facts and feelings for the purposes of achieving a synergistic decision-making process. The outcome may be positive, negative, neutral, or a combination, but the process of making decisions is more fully informed with both the mind and heart at work. For me, this is the pragmatic value of intuition.

One of the frequent expectations of executives is to lead an organization forward, into the unknown. Executives must prognosticate in order to optimally position the organization for what is to come. By definition, the facts of that future are incomplete; otherwise, repositioning would be formulaic. Likewise, the inability to observe emergent statistics, incorporate predictive trends, and acknowledge environmental complexities would leave an organization subject to an evolving context without the benefit of knowing that you are making decisions to move the organization in a concerted direction. Organizations also exist to further human interests, and executives are expected to be mindful of doing the right thing—that is, to look beyond the bottom line (profits) to also include the responsible stewardship of people and the planet. Can the executive mind alone process so many factors to ensure effective decisions? Perhaps or perhaps not, but working without employing all the potential ways of knowing would seem to restrict the possibilities.

Maturity and experience sometimes brings with them wisdom. While knowledge appears to live in the mind, I have concluded that wisdom lives in the heart. As the dean of the College of Business and Management at Lynn University, I can see this every day when some aspirant executives in the MBA program report doing everything right but failing miserably, while other aspirant executives report, seemingly, all the wrong moves but end up rising in success. How could that be? It isn't that one executive was smarter than another, as, academically, the two groups norm similarly. It isn't that the tools and techniques of management science are irrelevant either, as both report using what they know. Rather, it's that those who are, on the surface, moving in illogical ways but still finding considerable success are employing their full capacity—mind and heart—in the making of decisions.

Like some of the aspirant but less successful MBA students with whom I have worked, early in my own career I was very driven for success—making all the right moves, connecting with the right individuals, and positioning myself in the right place, poised to seize the next opportunity as being of paramount importance. I lived in my mind. And yet, early in my career I found myself stymied by the inability to achieve my aspirations. What was wrong? As I lost traction, I seemed to also lose motivation, capacity, and competency. It became a downward spiral, culminating in the election to demote myself into academic obscurity from my administrative appointment, becoming that reclusive member of the faculty in the office far down the hall from whom few seek counsel.

It was in the abyss of academic irrelevance that something shifted. It was after desperately flailing for something on which to latch that I let go—let go of expectations and anticipations. I let go of the future and, for what seemed like the first time, I began to just breathe in the present moment. And in that breath, I became aware of a bigger space in which to exist. I can't say if it takes hitting the bottom to find your way back, but ever since that experience I have come to appreciate that one's journey, no matter how trying, is just that—a journey, to locate the heart. Over the years I have explored, both academically and experientially, what was revealed to me in that moment and have learned that much of what I discovered has been known to many before me and through many traditions. I had discovered nothing new, but I had discovered it for myself. From the language of the heart, I know it as wisdom. From the language of the mind, I know it as intuition, and in the latter capacity it can be a powerful tool for the executive looking for a more holistic approach to decision making.

CULTIVATING EXECUTIVE INTUITION

As a long-serving educator, and currently as dean of the College of Business and Management, I have firsthand familiarity with the emphasis of our Western educational system on the development of the mind to meet personal, organizational, and societal goals. We teach students that using our minds better means thinking more critically, rationally, and systematically—at least in the realm of business studies and, I could argue, in other disciplines as well. This educational approach is informed by the perspective to identify thinking as being or, as stated by the father of modern Western philosophy, Rene Descartes, "I think, therefore I am."

Decision making, then, is rooted in the diagnostic thinking practices of mentally organizing phenomena as they relate to the greater known context. This context is informed by our lived experiences, taught structures, and belief perspectives. As executives, we are expected to integrate factors of the decision-making process into existing thought frameworks for, presumably, rendering a well-thought-out decision. It is convergent thinking. We isolate factors, determine interdependent variables, and seek to define the problem. From there we can identify gaps requiring resolution and begin to seek solution options for consideration and to make conscious choices for action. The problem is that along the way our mental calculations can get hijacked.

WRANGLING CONVERGENT THINKING

When we approach decision making solely using our existing ways of knowing (experiences, structures, and perspectives), we not only delimit possibilities but often compound the problem with our unconscious conditioning. Unconscious conditioning shows up as evaluative self-talk and fragmented focus, leading to impulsive decisions that are, all too often, faulty. The thinking mind overthinks and conflates the decision with a myriad of irrelevant concerns. The first key to overcoming this tendency is to wrangle the judging mind.

Eastern educational systems have long been informed to focus on cultivating and developing *attention*. It was felt in days past that you were not fully developed as a mature person unless you had one of the arts mastered. It was understood that it was attention through mastery of an art form (broadly interpreted) that grounds a person from the perils of overthinking and conflation. While there are many routes for achieving this attentive state of existence, the common core is to have developed a contemplative practice. It is not mastery of the selected art form or particular practice, although this very well may occur, but the practice itself that becomes the honing of attention. Through a cultivated contemplative practice, one is able to *focus* attention on the present activity and learn how to disengage with that which distracts attention. We see every day the challenges of remaining focused in the tendency of many to multitask during important activities such as driving, studying, or meeting. In fact, we often see employers seeking candidates who are able to multitask as though this is a skill of value, but research has clearly demonstrated that without training on the ability to switch quickly between tasks, multitasking leads to fractured attention, resulting in delayed task accomplishment and poor outcomes, including that of decision making (Frick, 2015).

However, we often think of eliminating distractions only of the external form. When it comes to decision making, probably the most significant distraction has an internal origin. Baggage from the past and worries about the future conflate with the present decision to be made and unwittingly interfere with our ability to consider, unfettered, the decision variables and choice options. This is where contemplative practice has value. Through a mature contemplative practice, although improved results can often be observed more immediately, we learn to pause to recognize conflating factors, to disentangle mental distractions, and to acknowledge that we have other choices than those to which our minds may have immediately jumped. I like to think we learn to allow space for a breath to occur between what we perceive and how we react to that perception. We learn to not be impulsive in our thinking. In Eastern traditions, this is sometimes referred to as *calming the monkey mind*, and in the West it is referred to as *emotional regulation*. In either respect, it is the wrangling of the judging mind that disengages convergent thinking in the decision-making process. By doing so, we allow for something else to enter into the breathing space, and I contend that this allows for divergent thinking to occur. We gain choice in our response to the stimulus. It is in thinking divergently that we also allow the intuitive process to flourish.

CULTIVATING DIVERGENT THINKING

When we begin to hone the practice of recognizing our emotional responses to stimuli, we can also begin to ask the question, From where does that impulse originate? Asking the question alone allows for space to develop, and reflecting on its origin helps to recognize that it may come from a place that is unhelpful to making the present decision, such as baggage from the past or worries about the future. We immediately gain a choice in how we respond to accommodate our impulse: honor it as coming from a place of relevance or dismiss it as irrelevant to the present decision. Once we recognize that we can retake control of hijacked emotions, we become able to not only increase our options but also, importantly, elect to not react at all or at least for the moment. Doing so creates even more space for additional insight to be gained.

By allowing more space to develop, we engage in openness or the ability to hear our intuitive mind. I stated previously that I believe wisdom lives in the heart and has much value to add in rendering effective decisions for the executive. But that wisdom gets transmitted through the mind when choices are made and often forms into something other than a concrete or discrete form for conveyance to or understanding by others. Herein lies the reason why the cultivation of a contemplative practice is of paramount importance. When executives report that they "have a feeling" about something, it may come from two different perspectives. For the executive acting from a place of impulse, that feeling is belabored by baggage or worries, absent of consideration of their origin. Thus, that feeling is not being made from a place of choice, given the relevance of the feeling. For the executive acting from a place informed by contemplative practice, not only is that feeling informed by rational reflection on the sources of origin but, additionally and importantly, it is also informed after having cultivated space for the wisdom of the heart to translate into an intuition. If intuition was rational it could thus be explained, but the definition of intuition implies the ability to know something based on insight. It is only after cultivating space to allow for divergent thinking that the wisdom of the heart can penetrate our thinking to provide insights to inform decision making. When it does, that's intuition.

I believe that allowing space for divergent thinking wherein the heart–mind connection can take place is essential for executives seeking to realize value from one's full capacity. Of course, not all decisions allow for time, but honed practice enables the mind–heart connection and intuition to happen more organically to meet more immediate needs when it has been cultivated from consistent practice over time. Consistent contemplative practice developed over time builds on Noel Burch's skill development model (Adams, n.d.) and contributes toward an executive's capacity to move through the stages of unconscious incompetence (monkey mind), conscious incompetence (emotional regulation), conscious competence (beginner's mind), to unconscious competence (wisdom) with respect to listening to the heart and allowing the mind to interpret what is heard into intuition about a given situation and decision to be made.

GENUINENESS OF PURPOSE

Ultimately, trusting your intuition as an executive in decision making necessitates accepting an idea to which I refer to as *big mind*, or the ability to dispense of dichotomous thinking of mind or heart by integrating an embodied feeling for the whole of a situation. Bifurcation gives way to full sensory knowing, and we can come to trust it. But is that a risk an executive can accept?

I believe executives have risk in all decision making. If, as stated earlier, all was known regarding the circumstances of phenomena in advance, then no decision would need to be made as there would be predictable outcomes. Risk is inured to the work of executives. But an executive can become more comfortable with risk when trusting an embodied way of knowing the world in which they take part. In other words, executives gain a more solid footing by appreciating the greater good of our work and allowing a genuineness of purpose to serve as our guide. When we come to decision making with *right intent* and build from cultivated contemplative practice, we at least know that the spirit of decision-making engagement comes from a place wherein we can feel comfortable with the decision irrespective of the outcome. For me, I live more comfortably with the outcome by virtue of the process used in rendering the decision when coming from a place of honest intention and cultivated practice. When I've done the best that I could do, then I've done what I can do.

DECISION-MAKING CHALLENGES

Since that time of finding myself in academic obscurity, I have come a great distance with my practice and, I presume, the effectiveness with which I am able to rely on intuition for decision making in my role as a dean in higher education. Prior to my current role, I have served as a dean in two other capacities: first as dean for a school of continuing education and later as the dean for global education and special initiatives. In the former, I found myself helping an institution develop academic programs serving a rapidly evolving local economy, and in that experience, I would have my first awareness of the value of employing the intuitive decision-making processes outlined. The academy is traditionally slow and deliberate in its decision-making processes, but entrepreneurs, business leaders, and the marketplace can shift on a dime. As dean for a continuing education unit, it was imperative that I remain plugged in to local economic developments while also respecting the academic community to which I was contributing. Focusing attention on local developments, cultivating openness of inclusion with academic partners, and honoring a genuineness of purpose to advance higher learning were key to successfully keeping programs current, revenues generated, faculty inspired, and the expectations of students as well as community members met. Using an intuitive process resulted in positive outcomes for various stakeholders, the institution, and me.

It was that ability to balance diverse interests that led to my second appointment as dean, focusing on global education and special initiatives, wherein I was able to really come to terms with the power of an intuitive decision-making process. In this latter role, I was again asked to bring together interests across a diverse institution to identify an inspiring project on which to focus the future growth of the institution (special initiative). Working for an institution with regional accreditation, we had requirements to meet in terms of compliance that focused on past performance. But a new change to the accreditation process required institutions to develop a second initiative that focused on the future of the institution, derived from a deliberative and inclusive process as well as one that engaged the institution in an area in which it had not previously engaged while having broad institutional impact. The institution for which I served as dean was one of the very first institutions on which this requirement was placed as a new accreditation standard. We were entering uncharted territory.

The context of leading change into an unknown is what gave power to the processes I have outlined in this chapter. The challenge was set to me and I readily accepted without knowing just how it would be accomplished. However, it was the ability to focus attention, cultivate openness, and keep genuineness of purpose forefront that provided the structure to focus an institutional community on moving forward without a roadmap for how it should be done. Leading this task, I was able to strike the tone of a facilitator to the process, helping less with making a decision and more with guiding a decision-making process. In fact, it was the divestiture of personal interests in the outcome that enabled the full power of intuitive tools to be deployed for the benefit of the institutional decision-making process. Wherein using the intuiting process I have described gave me the capacity in my first role as dean for individual decision making, the intuiting process was even more powerful when utilized by an institutional community to render decisions on how to proceed and on what ultimately would become the future thrust of the institution. In the end, we not only met the expectations of accreditation but also gained considerable attention for the early success and impact on student learning. Most importantly, it was culture shifting for the institution and became a guidepost for future institutional decision making.

Now that I serve as dean for the College of Business and Management, the process of honoring intuition is informing how I elect to move forward with decisions impacting the college. As with any executive, I seek to lead my organization forward to successfully realize an optimal future. I have come to trust my intuiting process and know that it will continue to be of support to me, pending that I continue to focus my attention, cultivate openness, and remain true to my genuineness of purpose as an educator and executive.

CONCLUSION

From my perspective, intuition lives neither in the gut nor the hip but rather in the heart. More so, it serves us best when integrated into the holistic thinking allowed by the space created when we focus our attention, cultivate openness, and remain true to our genuineness of purpose. As such, it does not happen in a vacuum but is part of an executive using full and broad capacities in rendering a decision. Finally, using intuition effectively in executive decision making is the result of having honed a contemplative practice that creates space for choice and is open to new insights that speak to the heart–mind connection. I hope in sharing my experience and perspective I have succeeded in providing something of value for others to consider in their work as executives and in making decisions. It may not be the only way for an executive to cultivate or use intuition, but for this executive and educator it has proven worthwhile.

REFERENCES

Adams, L. (n.d.). Learning a new skill is easier said than done. Gordon Training International. Retrieved from http://www.gordontraining.com/free-workplace-articles/learning-a-new-skill-is-easier-said-than-done.

Frick, W. (2015). The curious science of when multitasking works. *Harvard Business Review*, January 6. Retrieved from https://hbr.org/2015/01/the-curious-science-of-when-multitasking-works.

Harwell, D. (2018). A down day on the markets? Analysts say blame the machines. *Washington Post*, February 6. Retrieved from https://www.washingtonpost.com/news/the-switch/wp/2018/02/06/algorithms-just-made-a-couple-crazy-trading-days-that-much-crazier/?utm_term=.1173c3a6a011.

Mullen, M. and Ackerman, E. (2018). Can veterans rescue Congress from its partisan paralysis? *USA Today*, February 7. Retrieved from https://www.usatoday.com/story/opinion/2018/02/07/veterans-can-rescue-congress-partisan-paralysis-mike-mullen-elliott-ackerman-column/309188002.

7

Leadership Intuition Meets the Future of Work

Written by: David Dye

Contributing Writers: Chiara Corso, Claradith Landry, Jennifer Rompre, Kyle Sandell, and William Tanner

CONTENTS

"We're gonna be in the Hudson," the pilot said to air traffic control.

The weather in New York on January 15, 2009 was pleasant, but the inside of the cockpit of US Airways Flight 1549 was anything but. As the plane plummeted toward the river, the automated proximity warning system chanted, Pull up! Pull up! Pull up!

> "Got flaps out," said the co-pilot. "You want more?"
> Pull up! Pull up! Pull up!
> "No, let's stay at two," the pilot replied.
> Caution, terrain! Pull up! Pull up! Pull up!
> "This is the captain," he announced over the intercom. "Brace for impact."
> Terrain! Terrain! Pull up! Pull up!
> He turned to his co-pilot. "Got any ideas?"
> Pull up! Pull up! Pull up!

"Actually not," the co-pilot replied.
Pull up! Pull up! Pull up!
"We're gonna brace."[1]

***

As Chesley "Sully" Sullenberger lifted off from LaGuardia Airport in an Airbus A320, he had no idea that minutes later he would be hailed as a hero and a model for quick decision making and performance under pressure. Most of us know the story from news articles or the feature film depicting the event: against all odds, the brave airline captain manages to perform an emergency ditching of a commercial flight in the Hudson River after the plane struck a large flock of geese. The end result: all 155 aboard the flight survive, and passengers and crew will forever owe their lives to the swift problem solving of Sullenberger and co-pilot Jeff Skiles. Despite the fact that the event took a little under 3 minutes from engine failure to emergency landing, Sullenberger managed to draw on a lifetime of experiences to guide the plane to safety that day.[2] How did Sullenberger's nearly 40 years of experience as a commercial airline pilot combine with his assessment of the situation and his own feelings and values to form the controlled actions that he exhibited that January afternoon? As Sully put it, "One way of looking at this might be that, for 42 years, I've been making small regular deposits in this bank of experience: education and training. And on January 15, the balance was sufficient so that I could make a very large withdrawal."

Though certainly not as dramatic as Sullenberger's experience navigating the skies, the stakes facing leaders today have never been higher. As organizations, industries, and society at large evolve, so too does leadership at the executive level. The foundation and characteristics of great leadership remain largely unchanged, but the speed and urgency of decision making has come to the forefront as the single greatest factor impacting executives today. Industry disruption due to digital transformation, an evolving workforce, and other leadership demands is at an all-time high. Recent history is littered with examples of leading companies that have faltered due to their inability to act quickly and decisively. Additionally, leaders are now inundated with data to sort through and experts to consult before coming to key decisions that set the course for their organizations. Given the speed and complexity of decision making today, executives must rely on an age-old method: intuition.

We have all heard the advice to "follow your gut" before, but what does this mean? Gut-driven decisions are difficult to quantify and evaluate and as such are rightfully treated with a heavy dose of skepticism by others. However, there is a difference between the common "gut" platitude and what I am advocating as informed intuition. Executive intuition is an informed process, perfect for today's ever-changing business environment, because it blends the years of experience and insights leaders have cultivated with the powerful analytics tools that most organizations have at their disposal. Informed intuition is defined as the process of blending existing information and data with one's experiences, educated assumptions, and instincts to arrive at a logical conclusion. When viewing intuition as being driven by both instinct and insight, leaders can more easily integrate and make sense of complex information, resulting in decisions that are well-rounded, delivered with conviction, and appreciated by those who are impacted.

Intuition is best used during times of high-speed change, when the situation is ambiguous and risks are largely unknown. If that sounds familiar, it should. The modern business reality demands some level of intuition, and in this chapter, I make a case for the power and relevance of informed intuition. First, I will provide an overview of the changes to the business landscape that necessitate this new way of thinking. Next, I will review a framework for leadership decision making and show its connections to the development of informed intuition. Last, I will discuss how leaders can blend intuition with more rational, data-driven approaches to decision making, demonstrating how they can successfully capitalize on their intuition as one piece of a larger puzzle.

SHIFTING SOCIETY, SHIFTING LEADERSHIP NEEDS

It is more important than ever before for leaders to be able to exercise agility, adaptability, and other characteristics of informed intuition. Both the workplace and marketplace have rapidly shifted: technology continues to advance and evolve; the workforce is changing with the advents of millennial talent and artificial intelligence; and the phrase "Innovate or die" is increasing in salience as most every industry continues to mature within this "Digital Age."[3]

The very nature of intuitive decision making has changed with the evolution of big data and business analytics. Our traditional understanding of "following your gut" meant making decisions in the absence of data.

Today, however, leaders are inundated with data. Though more data can result in more reasoned approaches to solving organizational issues, the sheer volume, velocity, and variety of data can make it difficult for leaders to utilize properly. Not only do leaders have a plethora of descriptive and diagnostic analytics to guide their understanding of a topic, but advances in data analytics have now given business leaders predictive and prescriptive insights to take into consideration. These future-focused analytics allow leaders to accurately model different decisions and their potential consequences across the business, enabling a more reasoned, logical approach to situations that were once very ambiguous.

In order to succeed—even just to survive—leaders need to act as change champions for the Digital Age. Resistance to change isn't an option; a study featured in the Harvard Business Review found that CEOs who failed to adapt to the Digital Age were "being phased out" and showed that, in a survey of executives, the percentage of respondents who said their CEO was a "champion for digital" doubled over the past decade.[4] It is the new norm for leaders to prioritize digital innovation, at least in talking the talk, but exceptional leaders must also be able to walk the walk. This means not only supporting digital transformation within their organization but also having the expertise, confidence, and mindset to enable their organization to actively drive change.

Another writer for the Harvard Business Review describes this as "choreographing the change" instead of existing in a constant state of ad hoc reaction and contends that "only the CEO has the power to provide this kind of direction" for their organization to accomplish this.[5] Leaders must have a comprehensive understanding of the digital landscape that their organization plays in and a strong Digital IQ. As executives, they must also be able to bring a holistic point of view to the digital space; they cannot make decisions in a vacuum but must reconcile the complexities and contradictions of their entire ecosystem. This, again, underscores the need for strong informed intuition: inherently, when you bring all of these disparate pieces of information together from the ecosystem—the digital lens, your organization's mission, the fluctuating demands of the marketplace, workforce issues, and so on—you are using your own unique perspective in order to make sense of it all.

Increasingly, we now have to go through this process of sense making at breakneck speed. In the Digital Age, leaders are called on to make decisions at a much faster pace and in much more turbulent environments. As a result, achievement necessitates a high level of risk taking. Overall, the capacity for risk-taking experimentation—and for quickly scaling up experiments—is increasing in value.[6,7] And for good reason: according to research by Andrea Derler, Anthony Abbatiello, and Stacia Garr, organizations that embrace risk taking are "five times more likely to anticipate change and respond to it effectively and efficiently, and seven times more likely to innovate than others that do not."[8] Informed intuition is critical to thriving as a leader in these types of environments.

In the face of such a wealth of data—both descriptive and prescriptive—where does informed intuition fit? And how do leaders use their intuition to make smart decisions on what types of risks to take? Leaders are faced with ambiguity and complexity in the uncharted terrain of the Digital Age, while being simultaneously inundated with data. First, leaders need to get comfortable embracing the unknown—enjoying the breeze on their face as they move from leap of faith to leap of faith. For direction on the practical application of this insight, look no further than Jeff Bezos, who suggests that executives need to make decisions with 70% of the data. Affirming my earlier point about the need for speed in the Digital Age, Bezos cautions, "If you wait for 90%, in most cases, you're probably being slow."[9] Intuition can still play a key role as an additional input in situations when the data is simply not available or only tangentially related to the problem at hand. In such moments, leaders must draw from their personal experiences and insights, using these tools to correctly prioritize and simplify the existing data, or to communicate the most appropriate course of action in the absence of data. Thus, in today's fast-paced business world, the traditional idea of intuition or following your gut has fundamentally changed, necessitating the new ability to quickly digest information while blending it with one's experiences, educated assumptions, and instincts to arrive at a logical conclusion.

Regardless of how much insight they have on an issue, executives must strike a balance between engaging with the data streams and mechanical aspects of an issue and acting flexibly to respond to issues that resist a cut-and-dry solution. The growing need for this balance is affirmed in Deloitte workforce analyses such as "The Future of Work," where we identify this skill as "STEMpathetic talent."[10] Research suggests that the need for this type of intuitive ability will only increase as technological augmentation and artificial intelligence are used more widely to automate tasks. Organizations should expect a heightened need for leadership that can relate intelligently to the entire ecosystem in front of them, drawing on emotional intelligence, creativity, and social skills while still engaging technical knowledge and data analysis.[11,12] Other studies have supported this as well; as the world gets more digital, leaders must double down on the "human experience" in order to raise their Digital IQ. In the survey of executives I mentioned earlier, the top performers—those with revenue growth and profit margin increases (or expected increases) exceeding 5% for both the past and upcoming 3 years—were found to have a more developed understanding of how to center human experience in their digital endeavors.[13] Even data experts agree with this approach. When asked about the characteristics that make a good data scientist, Rachel Schutt, a senior statistician for Google Research, described a person with "deep, wide-ranging curiosity, [who] is innovative and is guided by experience as well as data."[14]

A FRAMEWORK FOR DEVELOPING INFORMED INTUITION

The question of how organizations grow these types of curious, emotionally intelligent, innovative, and data-fluent leaders is one that we've considered thoroughly during my time at Deloitte. Our leadership framework in Figure 7.1 provides an overlay onto the concept of informed intuition. The Deloitte Leadership Framework identifies the three key components of a leader: capability, potential, and experience. In terms of capability, leaders must be able to inspire others, achieve results, exert influence, collaborate, provide vision and direction, show business acumen, bring a competitive edge, and build talent. While many of these capabilities are developed through practice, many others require some innate drive or personality dimension. For example, while you can receive coaching on your rhetoric, demeanor, and actions to improve your ability to inspire others, your personality and charisma may dictate the extent to which that training can have an effect (Figure 7.1).

This leads us to the second component: potential. In our framework, we define potential as a leader's ability to drive and respond to change, to think quickly and flexibly, to adapt to complex interpersonal or emotional demands, and to intrinsically and extrinsically generate motivation to achieve. Personal values are also encompassed within this component. It's

FIGURE 7.1
Deloitte leadership framework (from https://www2.deloitte.com/content/dam/Deloitte/at/Documents/consulting/leadership-booklet.pdf).

clear to see how closely the dimensions of potential align with the needs of a rapidly changing Digital Age. In both leading researchers' analyses of the Digital Age and our Deloitte Leadership Framework, adaptability is paramount for success.

If we think of our main ingredients for informed intuition as facts, feelings, and values, there are clear connections to the Deloitte Leadership Framework. This framework demystifies some of what goes into informed intuition, and perhaps more importantly, it helps organizations develop a pipeline of future leaders who can capably and confidently wield intuition to make innovative and game-changing decisions. The Deloitte Leadership Framework is data driven itself, using assessment and diagnostic tools to drive leadership strategies for a diverse customer base. It's also executed through cognitive modalities and engages clients through holistic interviews.

At the center of this framework is the idea that the organization or ecosystem—not the individual him or herself—has the biggest influence when it comes to leadership development. The data, the situation, and the environment all matter. But if you don't have control over your entire organization, what are some ways to improve your immediate surroundings to strengthen informed intuition? For one, you can build a diverse team that can complement your capabilities, potential, and experiences. In the Digital Age, it is critical to admit when you don't have all of the answers. A tactical approach is necessary to draw on a team that can bring vastly different perspectives to a difficult problem that demands a quick response. Second, you can intentionally expose yourself to new situations to build a diversity of experience within yourself. Repeated problem solving can bolster your capabilities, but problem solving in a new domain can stretch and exercise the dimension of your potential. You can also dive into the Digital Age itself, blurring the lines between typical dichotomies that separate the technological from the emotional: using virtual reality to experience new situations, for example, or tapping into artificial intelligence that has been developed to use emotional reasoning. These are all examples of how you might develop your intuition, but what does the practical application of informed intuition really look like, and what real-world stories can we look to for inspiration?

***

As soon as Sully Sullenberger realized what had happened, he went into action. He knew that clear communication with his co-pilot would be critical to surviving the ordeal. He also knew that, being the captain, he was responsible. Sullenberger's first words after the plane's engines failed were not a lamentation, he simply said to his co-pilot, "My aircraft." His co-pilot, Jeff Skiles, recognized Sullenberger's authority and said back, "Your aircraft."[15] In this simple way, Sullenberger showed his capability by exerting influence on his co-pilot and in the situation at large.

After this simple exchange and understanding, Sullenberger made a decision that was unorthodox but perfectly describes his potential in thinking outside the box. Sullenberger flipped on the Auxiliary Power Unit (APU), a small turbine that supplies energy to critical systems such as hydraulics, pressurization, and electrical power in the event of an engine failure.[16] It is not the fact that Sullenberger turned on the APU that is impressive but rather when he did it: right at the start of the event. The reason Sullenberger immediately turned on the APU aligns with the third component of the Deloitte Leadership Framework: experience. Years prior, Sullenberger had experienced a double engine blowout while flying fighter jets in the Air Force. The jet lost electricity, and it was this experience of flying without electrical power that led Sullenberger to flip on the APU far earlier than what any checklist would have told him to do. The National Transportation Safety Board later verified the importance of turning on the APU at this critical juncture, stating:

> Starting the APU early in the accident sequence proved to be critical because it improved the outcome of the ditching by ensuring that electrical power was available to the airplane. Further, if the captain had not started the APU, the airplane would not have remained in normal law mode. This critical step would not have been completed if the flight crew had simply followed the order of the items in the checklist.[17]

In just 3 short minutes, Sullenberger had shown examples of capability, potential, and experience, allowing his intuition to guide him toward an emergency landing.

INFORMED INTUITION IN PRACTICE

Let us transport back to the office now. Using intuition can be an intimidating prospect for organizational leaders, especially those who are used to making primarily data-driven decisions and who may not be as used to tapping into their curiosity, innovation, and experience. Ralph Larsen, the longtime CEO of Johnson & Johnson, gave voice to this concern, saying, "Very often, people will do a brilliant job up through the middle management levels, where it's very heavily quantitative in terms of the decision-making. But then they reach senior management, where the problems get more complex and ambiguous, and we discover that their judgment or intuition is not what it should be."[18] What Larsen described is a common frustration for many organizations, as critical succession planning efforts can fail due to a promising leader's inability to tackle difficult business problems using an intuitive way of thinking.

Earlier in my life, I had a career-changing experience in which my informed intuition played a significant role—in my decision, in my career, and now how I look back on it. In essence, it was the seminal moment in my career when I transitioned from being a technical expert to an executive leader. My boss brought me into her office and said, "I would like you to take on this new exciting role that will change the way we do business around here. The role comes with risk and uncertainty and it also may not be very popular with our customers in the short term. But when I look at who you are and your potential as a leader, I have complete faith in your personal success and of our organization." I will never forget what she said next: "If you succeed, I will sing your highest praises. If you stumble, I will be there to catch you when you fall." In my decision to accept the role (very quickly I might add), I certainly considered the facts and my feelings. But when I look back many years later, I truly believe my values played a role in not just accepting the assignment but in articulating and convincing others to rally around the new organization. For me, this life-changing event changed the course of my career; more importantly, it shaped how I view myself as a leader today and how I look to provide others with similar opportunities. My informed intuition as a brand new executive was certainly at play.

Part of the problem is that using your intuition is scary, especially if you subscribe to the old view of intuition as a gut decision devoid of reason. However, as the scientific community learns more about intuition, we are coming to understand that it is instead a remarkable process by which we unconsciously leverage our past experiences in confronting the problems of the present. Brene Brown explains the concept well:

> Intuition is not independent of any reasoning process. In fact, psychologists believe that intuition is a rapid-fire, unconscious associating process—like a mental puzzle. The brain makes an observation, scans its files, and matches the observation with existing memories, knowledge, and experiences. Once it puts together a series of matches, we get a "gut" on what we've observed.[19]

Leaders using informed intuition take it a step further by understanding how their memories, knowledge, and experiences led them to a particular decision, and properly weighing their intuition amid the myriad of relevant information when deciding. However, using informed intuition to come to a decision is only the first step for leaders at the executive level; the next step is executing on intuition by convincing others to follow.

There are two things leaders must do to execute on intuition: first, they must have conviction that their decision is the right one; second (and related), they must be able to understand and articulate how they came to their decision. Intuition without conviction will leave people unconvinced and will stall progress, but having conviction in intuitively driven decisions is difficult due to the level of uncertainty that is inherently involved. Uncertainty is a natural reaction to intuitive thinking, but the best leaders recognize this uncertainty while also trusting in themselves and the many ways they have developed knowledge and insight over the years. Kathryn Finney, founder and managing director of digitalundivided, showed a great deal of conviction when she invested $30,000 of her own money toward a startup tech conference geared toward young women of color in 2012. As Finney explained:

> Most people think leaders are some sort of "all-knowing-beings," but in reality we often don't know the answers. The difference is that leaders trust their instincts to lead them to an answer. [...] No one had ever done a startup conference for black people before and there was really no model for us to follow. People thought I was crazy and told me so. But I did know that there was a disconnect between the tech startup world and my community.[20]

Despite a previously failed business under her belt, Finney held the conviction that her new idea would land. Six years later, Finney boasts that digitalundivided has helped build 52 companies, raised $25 million in investments, and reached 2000 founders.[21]

In addition to conviction, leaders must possess the self-awareness to understand and articulate what inputs played a role in their decision-making process. Executive-level decisions have the potential to impact thousands of employees and customers. Due to the weight of executive-level decisions, as well as the rise of participative decision making, it is essential that leaders share the details of their decision-making process. However, developing the insight needed to understand the inputs leading to a critical decision and relaying such quick and complex thought processes to others is often easier said than done.

A common example comes to us from the world of sports. Professional athletes are often asked after making a game-winning shot or goal to "take us through the moment" and describe what happened. Very rarely do these athletes point to the myriad of inputs that led to their decisions in the moment, and how could they? These athletes' split-second reactions were likely based on years of experience in similar situations, advice from coaches, and other inputs that snowballed to create a quick, fluid reaction. However, upon further reflection, they may point to a film session that showed their opponents' tendencies, or compare the situation with a similar one from a previous experience. Oftentimes, they point to their coaches, who put them in a position to succeed. These inputs combined to put them one step ahead of the competition, and the best athletes not only draw from diverse sets of information very quickly, they can later identify the factors that led them to success, with an added benefit being that they can share with others or keep drawing from the most helpful sources.

There may be no better example of this quick learning process than hockey superstar Sidney Crosby. Crosby is renowned not as the fastest player nor the player with the most accurate, hardest shot, but rather as a player who can quickly learn and implement a new strategy that puts him ahead of the competition.

The business world is not much different in terms of speed of decision making. Executive decision making is often just as quick as a hockey player's decision to shoot or pass, employing what psychologist Gerd Gigerenzer termed "fast-and-frugal thinking"—decisions we must make in the blink of an eye when time, knowledge, and/or cognitive resources are limited.[22] Where the business world differs, however, is the level of complexity of the information and the need to draw from multiple areas outside of one's specialty. To succeed as leaders, modern-day executives must develop the capacity to quickly interpret their intuition as one more input within an entire ecosystem of information. By reaching this standard of informed intuition, business leaders can provide a distinct competitive advantage for their organizations.

Former AOL CEO Robert Pittman provides an excellent example of a leader who has flourished in the era of complex, fast-moving data. As Pittman explains: "Staring at market data is like looking at a jigsaw puzzle. You have to figure out what the picture is. What does it all mean? It's not just a bunch of data. There's a message in there." Pittman's answer to the most complex of problems is, paradoxically, to seek out more data. He further explains: "Every time I get another data point, I've added another piece to the jigsaw puzzle, and I'm closer to seeing the answer. And then, one day, the overall picture suddenly comes to me."[23] What may seem like a sudden strike of intuition is actually a careful consideration of many factors and the development of associations between disparate data points, all happening behind the scenes. Pittman's intuition is actually far from a sudden inspiration, though it may seem that way.

***

When Sully said that he had "been making small, regular deposits in this bank of experience," he didn't mean that those deposits were singular or isolated. Rather, like compounding interest, those deposits multiplied by the number of connections he made between those experiences. Our informed intuition includes how quickly we make connections between our vast life experiences and the current situation.

Sullenberger's assessment of the situation certainly rings true, but the speed in which he accessed this fountain of experience and his ability to make connections in a completely novel situation highlights the incredible power of informed intuition. The fact that he worked closely with his co-pilot to assess options and settle on an answer further displays his ability to draw from multiple sources of data in arriving at a conclusion. Last, Sullenberger showed conviction in his decisions (as well as his values) by being the final person off the plane as it slowly took on water in the Hudson River.

Informed intuition is a powerful yet often overlooked tool that is becoming increasingly important in the new, digital era of work. Contrary to common belief, many major decisions cannot be mapped using big data or predictive analysis; they are simply tools to be leveraged when coming to a conclusion. Similarly, the traditional view of intuition as a gut feeling devoid of any deeper understanding is also limiting. What I am advocating is a differentiating approach to decision making whereby leaders blend data-driven methods with their established intuition to arrive at big-picture answers to their businesses' most challenging problems. In this manner, leaders are using the information at hand as well as their established intuition that has been built through experience, and instinctively understanding when and how to leverage each.

HOW TO DEVELOP INFORMED INTUITION

The journey to using intuition can be difficult, but there are steps leaders can take to drive these dual modes of thinking and arrive at informed intuition.

- First, leaders must learn to combine hard data and emotional skills. I am not advocating that leaders should disavow a rational, data-driven approach to decision making. In fact, leaders should be actively engaging with analyses and strategically using relevant data streams. If you are not capitalizing on the incredible advancements in analytics and big data, you are already behind. However, it is also true that leaders will not succeed if they arrive at conclusions based solely on preexisting data. As leaders rise in the ranks, decisions become broader, more complex, and more ambiguous. In such scenarios, relying on intuition as an additional input is advisable. However, intuition without emotional skills such as self-awareness makes it difficult to articulate to others that your decision truly is the best way forward. Thus, a strategic combination of hard data and emotional skills is a competitive advantage for any leader.

- Second, leaders must learn to embrace risk, as using intuition is ultimately an exercise in risk tolerance. Intuition is commonly used to "fill the gaps" in areas where the data is not available quickly enough or is insufficient to the matter at hand. Thus, with the use of intuition comes a certain amount of uncertainty. Leaders will not have any charts or figures to point to when describing their decision-making process, but must trust that their accumulation of experiences, organized and turned into wisdom, is accurately guiding the process. As such, leaders must have conviction that their experiences matter and they must be able to deliver their recommendations and calls to action with confidence.

- Third, leaders must continue to adapt quickly. It is not an exaggeration to say that leadership is all about adaptation and experimentation. What works in one scenario may not work in another because, at the core of it, people drive decisions. As Isaac Newton bemoaned after losing his life savings in investments, "I can calculate the motions of the heavenly bodies, but not the madness of people." Leaders today are in the same predicament Newton was in so long ago, as they find themselves operating in complex global markets with many unpredictable elements. There are no one-size-fits-all solutions. But, thankfully, informed intuition is set up for this level of ambiguity. Intuition is made up of associations between related concepts and events, and can be honed by making decisions in ambiguous environments where data cannot provide a quantitative answer. If leaders start small, they can begin building associations between more and more data points, eventually preparing themselves to employ informed intuition in the most complex environments.

- Fourth, leaders should check their intuition with others. With only seconds left before the plane hit the water, Sully asked his co-pilot, "Any ideas?" He was checking his decision making one more time before he committed to the action. He was testing his intuition against the logic and insights of his fellow pilot. The response from Jeff Skiles of "Actually not" was the confirmation that his intuition was right. This act of checking with others to verify your intuition is not relinquishing your authority but a tactic for gaining buy-in and commitment.

Bruce Henderson, founder of the Boston Consulting Group, defined intuition as "the subconscious integration of all the experiences, conditioning, and knowledge of a lifetime, including the cultural and emotional biases of that lifetime."[24] Henderson's definition from 1977 is still relevant, as intuition alone is still subject to bias, but he also could not have predicted the data, tools, and resources available to modern-day leaders. Leaders today can balance the inherent shortfalls of our traditional notions of intuition by considering it as one input to be analyzed in conjunction with other pieces of information. Thus, the concept of informed intuition blends the best of both worlds—rational and intuitive—resulting in decisions that can keep up with the rapid pace of the modern business world.

ACKNOWLEDGMENTS

The authors would like to thank Dick Richardson (Co-founder of Experience to Lead and author of Experience Apollo: Leadership Lessons from the Space Race) for his contributions to the chapter and for helping to bring the story of Sully Sullenberger to life.

NOTES

1. https://www.ntsb.gov/investigations/AccidentReports/Reports/AAR1003.pdf
2. http://www.nj.com/hudson/index.ssf/2009/01/timeline_released_of_us_airway.html
3. https://www2.deloitte.com/content/dam/insights/us/articles/4322_Forces-of-change_FoW/DI_Forces-of-change_FoW.pdf
4. https://hbr.org/2017/05/how-the-meaning-of-digital-transformation-has-evolved
5. https://hbr.org/2017/01/to-lead-a-digital-transformation-ceos-must-prioritize
6. https://www2.deloitte.com/content/dam/Deloitte/us/Documents/public-sector/us-fed-digital-organizations.pdf

7. https://www2.deloitte.com/insights/us/en/focus/human-capital-trends/2017/developing-digital-leaders.html

8. https://www2.deloitte.com/insights/us/en/deloitte-review/issue-20/developing-leaders-networks-of-opportunities.html

9. http://www.businessinsider.com/jeff-bezos-explains-the-perfect-way-to-make-risky-business-decisions-2017-4

10. https://www2.deloitte.com/content/dam/insights/us/articles/4322_Forces-of-change_FoW/DI_Forces-of-change_FoW.pdf

11. https://www2.deloitte.com/content/dam/insights/us/articles/4322_Forces-of-change_FoW/DI_Forces-of-change_FoW.pdf

12. https://www2.deloitte.com/content/dam/insights/us/articles/4438_wikistrat-white-paper/4438_The%20evolution%20of%20work.pdf

13. https://hbr.org/2017/05/how-the-meaning-of-digital-transformation-has-evolved

14. http://www.nytimes.com/2012/12/30/technology/big-data-is-great-but-dont-forget-intuition.html

15. Interview with Dick Richardson.

16. https://aerospace.honeywell.com/en/blogs/2016/september/the-miracle-on-the-hudson

17. https://www.ntsb.gov/investigations/AccidentReports/Reports/AAR1003.pdf

18. https://hbr.org/2001/02/when-to-trust-your-gut

19. Brown, B. (2010). *The Gifts of Imperfection: Let Go of Wwho You Think You're Supposed to Be and Embrace Who You Are*. Centre City: MN, Hazelden.

20. https://www.fastcompany.com/3031922/10-women-in-leadership-share-their-secrets-to-success

21. https://www.digitalundivided.com/

22. Gigerenzer, G., & Goldstein, D. G. (1996). Reasoning the fast and frugal way: Models of bounded rationality. *Psychological Review, 103*, 650–669.

23. https://hbr.org/2001/02/when-to-trust-your-gut

24. https://hbr.org/2003/05/dont-trust-your-gut

8

Spiritual Management: A Personal Account of Executive Intuitive Awareness

Written by: Eugene W. Grant

Decisions, decisions, and more decisions. We all make them every day. We make them when we get up. We make them throughout the day and we make them before we go to bed. Making a decision can be agonizing, depending on the issue we have to decide. As leaders, we must consider the consequences of our decision. How will my decision impact something or someone? Who will or will not benefit from my decision? What are the financial implications of the decision I'm about to make?

Like anyone else, I too struggle occasionally when having to make and trust a decision. When making a decision, I hope that whatever decision I make is the best choice. Moreover, as a person in an executive leadership role and responsible for the social, cultural, economic, health, and safety of a community, I must make decisions that are immediate or long term. Those decisions have impacts on children, young adults, families, seasoned citizens, businesses, and more.

Many people rely on logic, reasoning, education, experience, research, and empirical evidence as tools in the decision-making process. Like anyone, I too have an appreciation for each one. Yet, for me, as a person who is also guided and directed spiritually, I cannot overlook intuition, convictions, morals, and ethics. Each of these are not necessarily subordinate to logic, reasoning, education, experience, research, and/or evidence. But I do realize that it is important to use the aforementioned in most cases when making a decision. Intuition is really an unexplainable phenomenon. Whereas logic, reasoning, education, experience, research, and/or empirical evidence can and should be acquired and used in most cases, intuition is a combination of past experiences, emotions, and a gut feeling, which cannot be acquired but evolves over time. You cannot read a book or go to college or university and obtain intuition. It is an occurrence that I strongly believe develops over time through a belief in a higher power and knowledge of the self, along with faith and hope.

Some would argue that intuition is too risky to rely on, that using intuition will not give you the expected outcome. It's too emotional in its application. However, we also know that using logic, reasoning, education, experience, research, and/or evidence does not necessarily prove that the intended outcome of that decision will be correct either. When a decision is required immediately and there is no time for research, the professional experience is limited, the required education for that particular subject is not personally met, and there is not enough information to use logic, one has to use their instinct or intuition or, as with me, a spiritual connection.

Inventors or new innovators, creators of new industry, artistic works, and more usually rely on an innate sense beyond normal human learning and comprehension. It is an intuitive sense that drives them to making a decision to do or not to do. Steve Jobs, although very smart in technology, started a movement that intuitively caused him to believe that his Apple invention would revolutionize the communications and telecommunications industry. Henry Ford believed his mass production of the automobile would change the way cars would be produced on a massive scale. His overarching seminal quote regarding the human mind is well known: "Whether you think you can or you think you can't—either way, you're right." In each of these examples and others, conventional wisdom, logic, education, and the like were opposed to their belief. It was their past experience in their industry, their emotional connection to their product, and a gut feeling that their new inventions would revolutionize their industry. They tapped into something within themselves that sprung up like bubbling water out of the ground. They did not let what others thought, what conventional wisdom revealed, and what bodies of knowledge researched take them off the course of the intuitive self.

Steve Jobs once stated; "Your time is limited, so don't waste it living someone else's life. Don't be trapped by dogma, which is living with the results of other people's thinking. Don't let the noise of others' opinions drown out your own inner voice. And most important, have the courage to follow your ear and intuition." This era's technological icon believed that his intuition and gut feeling, which he referred to as "your own inner voice," would transform the way we communicate with each other. If he had allowed the opinions of his colleagues, investors, and even staff to cause him to drown in the ocean of doubt, antagonistic research, and the loud noise of naysayers, Apple, as we know it, would not exist. Thus, the world may not have developed the communications industry and social media model that dominates our societies today, which was built on Steve Jobs's intuition.

Sarah Breedlove, also known as Madam C. J. Walker, the first black self-made millionaire in America, must have had some form of intuition to guide her on the path toward entrepreneurial success during the era of Jim Crow. She earned money as a laundress with a salary of $1 per day. Even though she suffered skin and scalp ailments and early term baldness, her drive and determination led her to become one of the most successful African American hair care entrepreneurs in the early 1900s. Surely there was some force driving her through the trials and tribulations she suffered throughout her lifetime to propel her into the model of triumph she eventually became. Her take on her determination was, "I had to make my own living and my own opportunity. But I made it! Don't sit down and wait for the opportunities to come. Get up and make them." There were no other examples of successful black female millionaires, nor could you find blueprints in any library for building a multimillion-dollar company led by an African American. It was her gut feeling and her belief in herself and an understanding of her product that led her to positively alter the way black women cared for their hair. Her intuition taught her that there was an untapped market in the black community. That same intuition guided her to take advantage of it.

Planning a bold idea, a new course, and an unchartered direction is met with a level of trepidation along with excitement. It is considering the unknown future while believing and hoping that this new idea will work and be beneficial. It is taking a course that no one thought of and having faith and belief that you are chosen to bring it to life. It is intuitive.

Often for me, the invisible is visible. For me, faith and belief in Scripture, namely the Bible, is the intrinsic meaning of intuition. According to Scripture, faith is the substance of things hoped for and the evidence of things not seen. The inner voice is that substance or idea and the evidence that, while invisible to others, is visible within your own mind. Years of prayer, supplication, and fasting helps to develop the intuitive self. To others, their substance is based on tangible realities, whereas intuition is based on the intangible, which is not seen. Therefore, intuition relies on the internal machinations of the spirit mind and a relationship with a higher power, while logic and reason is based on the validity of external evidence and forces.

Many people looking at Seat Pleasant, Maryland, had low opinions and limited expectations. It was, in many people's minds, an undesirable city to live, raise a family, and age in place. Most developers overlooked Seat Pleasant, thinking that, if they developed here, their investment would not yield profitable returns. In addition, the community has many undesirable stores such as carry-outs, limited opportunities, high crime rates, double-digit unemployment, and a huge number of people living below the poverty level. The perceptions of crime caused many to fear coming into or even driving through the city.

Investors, developers, house hunters, and others would overlook our community and invest, build, and choose to live elsewhere. Federal, state, and county leaders had no vision and appeared to not even care about making improvements. Excuses of all kinds were made by these leaders as to why they did not reinvest federal, state, and county tax dollars in the community. As a result, the negative elements of our society saw an opportunity to prey, sell drugs, and commit crimes within the community. An environment was fostered that attracted these illicit activities, thus contributing to the damaging stigma of the city.

Even the residents held a negative view of their community. These residents were not engaged but continued to complain. The political leaders in office had limited or rather no vision for their community and never invested time, talent, or treasury into improving the conditions of the community. It was a city heading in the wrong direction. A place of hopelessness, doom, and gloom.

Along those lines, for many, Seat Pleasant had become the Invisible City, much like that in Ralph Ellison's monumental twentieth-century novel, *Invisible Man*. Our city was invisible. Her people, her assets, her features, and her beauty had all become invisible. Because of this invisibility, many had long ignored her, lost hope in her, and forgotten her. No one could ever see the potential of what she could ultimately become. Now, it must take a unique character, a special quality, and an unusual trait within the human mind to see the visible in the invisible.

To that end, in 2004, I became that person, the one with a vision for our city. Created by God and endowed by a unique spirit or power greater than myself and elected from among and by the people, I was chosen to lead. The people desired someone who could take them out of the ruins of antiquity to the apotheosis of hope, help, and healing. To help my city and to perform my duties, I often must rely on my intuition for the much larger and visionary ideas that will raise us to the lofty heights of our aspirations. While most decisions, particularly those day-to-day operational responsibilities, are **not** usually made intuitively, those broader and idealistic ideas, concepts, and mainly *visions* that I have come up with and act on are heavily influenced by intuition. It is seeing what is not seen. It is believing in something with no tangible evidence. It is moving toward a path where there are no past plans or best practices. It is working in a manner and direction that already exists, even though it doesn't. There is a confidence based on my belief that this higher power won't let me down, as it hasn't in the past.

Being in tune, on the same frequency, and in rhythm with this higher power requires prayer, supplication, and a laser focus. These three things are active in their use. You don't just sit and wait, daydream, and hope something will happen. It is taking time to really consider a situation and ask permission of this higher power to guide and direct you on a plain path. This plain path will lead you to the point of seeing the invisible. Then, there is a huge possibility that what you see, which no one else can, will come to fruition.

You must block out the loud noises of doubters who talk about every idea someone else comes up with besides their own. You must pull up the weeds of confusion that choke the life out of your idea. You must silence the voice of jealousy by being determined to work on the intuitive idea you have. Having intuition means that you are responsible for an idea that is ultimately good but that sometimes even good people will neither see nor except.

This level of intuition or faith is not magical nor mystical. There is no hocus pocus or abracadabra. It does not stem from a psychological disorder or dysfunction in the brain. Rather, it stems from an innate sense of self and a connection of that self to a positive higher power. That's when you believe you are surrounded by an optimistic energy that helps you develop an inner eye that has the ability to keenly see something that others may not be able to see, because it is invisible.

Focused leaders, inventors, innovators, and others like them seem to have a connection with a power that is greater than themselves. They seem to have the ability to see what others cannot readily see. Ford, Jobs, and Walker each in their own time and disciplines could see what others could not see. Before them, no one had been able to do what they had done, and now the entire world has been impacted by the intuition they relied so heavily on.

Likewise, I believe that I have been able to see what most others could not and cannot see about my City of Seat Pleasant. It is clear to me that

Seat Pleasant could be a vibrant place of endless possibilities. For me, I see a city critically transforming itself from just a low-income community with low educational attainment levels, high unemployment numbers, a high population of persons living below the poverty level, tremendous health disparities, no economic development activity in over 30 years, and diverted financial investment dollars. Yes, I see this community changing to a high-performing, digitally transformed and connected city where social justice is the platform that leads to inclusive and equitable economic prosperity. In this city, I can clearly see a place where we can educate our youth, improve our health outcomes, and create space for recreational and leisure activities that promote positive social interactions for all members of the human family.

Likewise, it is crystal clear to me that the flood gates of investment dollars will soon open wide and flow abundantly. One day, investors of all types will come and make real investments that will yield a profitable return and offer opportunities of employment to others. This much action will stabilize the city and reduce the levels of poverty and unemployment that currently stifle our city. This city will be a place where innovation can be created and tested, creating sort of a miniature Silicon Valley. It will be a place where ideas will be born and brought to the market for commercialization.

Because of the intuition and unyielding faith that I earnestly possess, we have begun that transformation. What I saw a few years ago is now currently in motion. Just a year ago, we went to Silicon Valley and signed an agreement with IBM, beginning our efforts in transforming Seat Pleasant into the world's first authentic Small Smart City. We began to outline how their Intelligent Operations Center would help move this city forward in its digital transformation. Meetings after meetings were held to develop the intricate details of the system that would help us deliver services to our residents that are faster, better, and personalized. We worked diligently to ensure that the solutions we were putting in place would help us to be more effective and efficient with taxpayer dollars.

However, this idea did not just happen. It began 5 years ago, stemming from a conversation I had at a conference with a very nice gentleman from IBM. He shared with me his work on *Smarter Buildings* and IBM's larger work on a *Smarter Planet*. The information he shared seemed to run parallel with my major project for the city. At the time, I was working on developing a 15-acre tract of land. This development would help shape my legacy. It was a net-zero renewable energy project with a power purchase agreement. It would have been the first of its kind in a small town. A new city hall, senior housing, a community center, a health and wellness center, and a state-of-the-art recreation center—15 acres of development off the grid and producing energy, creating a sustainable project that would benefit generations to come.

With great excitement I came back from the conference to Seat Pleasant with a sense of what we could do. The little bit of knowledge I acquired became like a seed that was planted in my fertile mind. But for others, I knew it would be a hard sell because their minds were not as fertile. They still believed in ancient and draconian methods. There would be those naysayers, those unbelievers and people who would find it difficult to till their minds to get prepared in understanding how we could change our community.

Over the years, I was confronted with many obstacles and had to endure many humiliating trials and tribulations. There was much vexation of spirit because of tumultuous attacks against my ideas. The forces of negativity confederated themselves to use whatever tactics possible to thwart my ideas. They did all they could to discourage me, foil my plans, and prevent from happening what I intuitively saw could and would materialize for our city. But I held on to my convictions because I had faith or intuition that by staying the course we would ultimately bring the invisible to the visible.

During this time of challenge and controversy, I had to pray and meditate. Often, I would grab my Bible and read various verses of Scripture to reinforce my belief and help me to not forget or lose faith in the intuitive nature of my vision. There were moments when I questioned when the paths of faith and foolishness would meet. Interestingly, those who have intuition sometimes struggle with the statement "There is a thin line between faith and foolishness." However, it is not foolishness. It is that deeply rooted belief that the connection to a higher power is revealing to you a vision that is plain to you but not to others. You come to realize that you are the custodian of this idea, this vision, and you are required to be the steward to bring it into reality.

There are times when the paramours of doubt and confusion conjoin to mount up and attack so fiercely that you almost give up and give in. But you don't. You remind yourself that in past experiences your intuition didn't fail you. You remember how you overcame adversities and obstacles. Thus, the cord of memory becomes long and wraps you like a blanket, providing you the comfort of knowing that your intuition is right. Therefore, you persevere. You keep going, you refuse to falter, and you never yield in the face of adversity. Doubt is never an option with the intuitive self. Your zeal and determination waxes stronger because the intuitive self knows what the end will be.

However, my enemies were successful. They did the unthinkable. Their actions were heinous. One night, they took a vote to terminate this multi-million-dollar project. This is a project that would have changed the course of our city. Their actions turned down 300 potential new jobs, a place where our seasoned citizens could move into and age in place gracefully, and a development that would have been the envy for the region. It would have given hope to so many who felt hopeless. Needless to say, I was devastated—crushed. It was incomprehensible. How could these people who say they love our city do such a horrible thing? How could they just turn their back on all the work that had been done over the previous 2 years? How could they live with themselves?

More importantly, I had to reconcile with the fact that this project would not happen. Night after night, day after day, I asked myself the questions: Was my intuition wrong? Did I misinterpret my intuition? It was puzzling to me, I was perplexed. Within me, my soul was in agonizing pain and I felt I had let down my community. However, I retreated to what I know best: prayer, meditation, and Scripture. After a while, I realized that I was not wrong. The interpretation of my intuition and my faith were stronger. Revelations became clear, causing me to work harder and not allow those who were fighting against progress to win.

So, I kept working, not on that project specifically because it was dead, but thinking, praying, meditating, and fasting more on the intuitive thought. As I continue my daily rituals and ponder on the vision, I realized that the specific project was not the solution, but more importantly my focus shifted to the outcome, which was more important. The more I focused on the outcome, the more my enemies worked.

During this period, people were planning for my destruction. They wanted to kill my spirit and force me out. They made numerous attempts to weaken me. They felt they knew what could cause me to quit. So, my enemies planned and were successful in taking my city-issued vehicle. They had written policies instructing staff not to take direction from me. They changed the hours of operation of City Hall, so I couldn't get into the building and continue my work. But the cherry on top was when they voted to remove the physical office of the mayor from the building, leaving me no place to work, meet with constituents, and conduct the official business of the city. This was a preposterous move that received national news and was an embarrassment to our city.

This embarrassment would make it even more difficult to sell our city and attract investments. Our residents were angry. We became an amusement for so many people. It was looking bleak. Yet, I was not deterred. My resolve had become stronger than ever before. Strangely, for me, it was confirmation that I was headed in the right direction. The vision I had for the city became even clearer, and I went to work without ceasing. As I embraced this better understanding of my intuitive self, the invisible became much more visible. Now was the time to forge ahead, be unstoppable, unmovable, and unshakeable. My faith required that I continue to press toward a mark with an even higher calling and purpose.

Henry Ford said, "When everything seems to be going against you, remember that the airplane takes off against the wind, not with it." The wind of evil people was against me. During those times, I experienced heavy turbulence. During the journey, I felt the effects of the winds tossing me from side to side. There were moments that I didn't know if I wanted to do this anymore.

Through this episode, I came to realize that the attacks were not just about me. This was not my battle nor my fight. It was against a greater purpose and plan that could ultimately unite and help a community. It was against the source from whence my intuitive self derived its ideas.

Yes, I realized that if I had been given the vision, then the source of energy and power that is greater than me would provide. Scripture revealed that my enemy would be my footstool, that I had to remember that I was not the tail but the head. My focus was on being the lender of ideas, not the borrower. It was clear to me that this higher power would make room for me and open doors that no enemy could ever close.

Through that dark period, I met a number of people from different walks of life and backgrounds. The intuitive self became more refined and the picture clearer. Since I didn't have an office, I was doing my work out of a local fast-food restaurant. The store owner was considerate and let me sit in the restaurant, set up my computer, and do my work on several occasions. One day, while sitting in the restaurant, I received a call from a gentleman in the governor's office. He asked me to join him for a meeting in Baltimore. He told me what the meeting was about, but it was still somewhat unclear to me. However, something inside me kept telling me to go.

The day came and I met the governor's representative in Baltimore. We went inside and it was a very small gathering. For some reason, I thought there would be a lot of people at this meeting. However, this small gathering of committed people shared information that helped me to better understand what my intuition had been revealing to me. It was like a confirmation. They began to speak about Innovation Villages and the benefits they brought to distressed communities. As I listened with great interest, a euphoric feeling swelled my spirit. It was excitement. My mind began to think of numerous ways this could benefit the people of Seat Pleasant. My prayers seemed to have been answered on what to do specifically. Once again, I was relying on my intuition, but this time I was more focused on the outcome and not necessarily the invention.

Once I returned to Seat Pleasant, I immediately went to work researching and learning more about Innovation Villages and the prospects of turning our city into one. However, my research revealed that there was already an innovation district in my county, and I realized that wouldn't work. So, I went back to what I know best: prayer, meditation, and scripture. As I went through my routine, the intuitive self showed me that we could be a Smart City with an Innovation Village. Eureka! There it was. The idea that I had with IBM combined with developing the 15 acres of land into an Innovation Village was the perfect fit. Remember, I read the intuition wrong; I originally thought it was about the project as opposed to the outcome. Sometimes projects/ideas work and sometimes they don't, but it's not about the project/idea for me; it's about the outcome. It was transforming our community and creating a better environment where people could live, work, play, pray, and stay. The intuitive self will always be in tune, in sync, and in rhythm with a higher power.

In reading Biblical Scripture, I read this passage: "Eyes hath not seen, nor ears heard, neither has it entered into the heart of man what God has in store for those who love Him." Well, I strongly believe this passage of scripture. While many may not see, cannot hear, nor will their heart receive this powerful intuitive energy, it is placed, however, in people who are connected to something greater than themselves, who believe in it and govern their lives accordingly. As a result, so many good things come about. The intuition of these visionary leaders forms the basis of major change.

For instance, Steve Jobs was a great visionary for Apple, while people within his company doubted him. Even more strange, many of his customers didn't believe in him at first either. His famous quote, "A lot of times, people don't know what they want until you show it to them." His ideas to transform his product were intuitively formed. When others didn't believe in him, ridiculed him, talked negatively about him, and even fired him from his company, he maintained his belief and faith in the intuitive self. Now, billions of people are impacted by his intuition.

Likewise, in 1914, Henry Ford became the first car manufacturer to use mass-produced parts in something as exact and at the same time challenging as an engine in an automobile. Ford had his detractors too. Yet, he strongly relied on his intuition, moving against the accepted norms and pushing forward to become the world's largest producer of affordable cars. The impact of his work is felt not only in the automobile industry but in many other industries as well. This is because of his belief and faith in the intuitive self.

Lastly, Madam C. J. Walker was a true entrepreneurial titan who developed a system of hair care for black women. In the early 1900s, no company had really developed a market in this industry for African American women. Because of her intuition, she saw what others could not see. She envisioned a company that specifically targeted an overlooked market. There were those who doubted she could be successful, given the opinions about black women at that time. But she did not allow them to cover her intuitive eyes. She worked hard, being relentless and training others, resulting in becoming the first self-made female African American millionaire. Because of her intuitive self and perseverance, the hair care industry reports that African American women spend approximately half a trillion dollars annually on hair care products.

While the story is still being written about Seat Pleasant, we are making progress. Our efforts to date have twice brought us before the United Nations to speak about our transformation and how our methods can be used to achieve the United Nations 17 Sustainable Development Goals. We presented at IBM's annual Think Conference in 2018, sharing our journey with over 40,000 delegates. We have participated in many conferences, workshops, and forums, sharing our journey and the importance of other small cities becoming "Smart."

All of this occurred as a result of me relying heavily on my intuition. For me, it is seeing the invisible as visible.

9

Creating an Intuitive Awareness for Executives

Written by: Ben Martz and Holly Morgan Frye

CONTENTS

INTRODUCTION

Most decision-making strategies have at their core a desire to ultimately choose the best or optimal answer. However, this optimal choice does not always occur. Early problem-solving researchers (Polya, 1957; Whiting, 1958; Osborn, 1963; Warfield, 1989; Warfield, 1990) identified four generalized problem-solving processes or activities: (1) discovery, the uncovering of information; (2) analysis, the decomposing of information into data and perspective; (3) synthesis, the recombining of data into information; and (4) choosing, the act of selection of a solution to the problem. Elements of intuition can be found in these activities. In fact, we suggest that intuition is "baked in" to many of today's leadership and decision-making theories.

INTUITION

To begin our discussion, we explore various definitions and explanations of intuition. The *Cambridge English Dictionary* defines intuition as "an ability to understand or know something without needing to think about it or use reason to discover it, or a feeling that shows this ability." The Merriam-Webster dictionary defines intuition with broad multiple strokes: (1) "quick and ready insight" and (2) "an immediate apprehension or cognition; knowledge or conviction gained by intuition; the power or faculty of attaining direct knowledge or cognition without evident rational thought and inference." For the purposes of this discussion, we are distinguishing between *innate* and *intuitive*. Using *Merriam-Webster* for comparison, innate is "existing in, belonging to, or determined by factors present in an individual from birth" or "originating in or derived from the mind or the constitution of the intellect rather than from experience." The small, subtle distinction "rather than from experience" between intuitive and innate gives us the basis for the remainder of our argument.

The intuition we are talking about here extrudes from a function of experience with the basic human problem-solving activities. When it comes to intuition for executives making decisions, we cannot ignore past experiences and how these experiences build up or contribute to the perceived intuition attributed to an executive decision maker. Continuing Zander et al.'s (2016) summarization that "intuition has been understood as an experience-based and gradual process," we can look to how experience helps build intuition. The first place to look should be how we, as humans, store experiences through memory.

When we talk of human memory, we naturally look to the vast research work and theories on understanding how individual memory works (Feigenbaum and Simon, 1984; Newell and Simon, 1972; Shannon and Weaver, 1949). In 1968, Richard Atkinson and Richard Shiffrin (1968), proposed the multi-store model for human memory as an enhancement to the common input–transformation–output model used at the time (Figure 9.1).

Briefly, our executive decision maker receives a stimulus, in the form of an observed problem or an insightful opportunity, from the environment requesting a fact or issue. That stimulus is translated and moves into short-term memory (STM), where, if a response is not immediately available, it

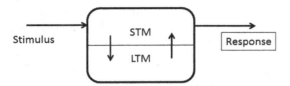

FIGURE 9.1
Generic human memory model with vanished state.

triggers a search request into long-term memory (LTM) for the retrieval of the appropriate information. LTM then supplies the retrieved information back to STM, where it can be used to enact a response. One of the vital activities necessary to make the LTM successful is to efficiently store data and retrieve information. One efficiency technique believed to be used by human LTM is to summarize large amounts of information into smaller chunks. In the field of human cognition and memory, one finds that researchers study how human memory deals with this summarizing characteristic.

Most people believe that learning is the process by which information is stored in LTM and recalled and applied in the future. At the physiological level, Saaty (2000) contends that memory is stored according to meaning. This follows the information systems model of data as symbols, while information is data in context. But this is not new: Piaget (1929) argued that the learning process and what is learned should be kept together as a unit. Cognitive researchers believe that the brain may combine related memories into more efficient structures in order to optimize recall and processing. These structures have taken on many names over the years. The concepts of scripts (Schank and Abelson, 1977), schema (Bartlett, 1932), schemata (Thorndyke and Hayes-Roth, 1979), templates (Sandelands et al., 1983), prototypes (Cherniak, 1984), and self-enacting response sequences (Roby, 1966) exemplify this area of thinking. These researchers argued that context represents the situation in which knowledge is acquired; therefore, they are permanently intertwined.

Gerd Gigerenzer (2007), a director at the Max Planck Institute for Human Development and author of the book *Gut Feelings*, suggests that "rules of thumb" have become a method of fast tracking decision making. The basis of this decision strategy is that it is easier to determine the information that is not necessary than that which will become the deciding factor. The underlying argument is that evolution in human decision making has created intuitive tools.

Two interesting implications for our discussion can be derived from this line of thinking: the concept of *implicit learning* and a plausible explanation for *déjà vu* (already seen). Reber (1993) explains that implicit learning is the acquisition of knowledge and understanding that cannot be ascribed to an explicit learning process. This idea is a little more subtle than knowing through experience where actual recallable knowledge and facts of past successes and failures in similar situations provides insight.

Researchers (Martz and Shepherd, 2003; Greenspoon, 1955; Jenkins, 1933; Thorndike and Rock, 1934) have used activities such as learning grammar, probabilistic learning, and matrix scanning to drive experiments on the concept of implicit learning. Subjects were asked to learn words and sentences with grammar rules from hypothetical languages. They were never explicitly told the rules of the language. Later, different but grammatically consistent words and sentences were provided to subjects. Implicit learning was measured by the extent to which a subject could identify the underlying grammar rules of the new sentences.

Piaget (1929) envisioned this implicit learning as succeeding through a process called *assimilation* and proposed it as the enabling learning process in early childhood development. In Papert's world, subjects developed models for problem solving from applying their current skills to the surrounding environment. The subject would then adapt their skills to enhance his or her solutions, thereby acquiring new, derivative skills. This process of using current skills within a problem environment to develop new skills is what Papert (1980) termed *appropriation*.

Other researchers (Schank and Abelson, 1977; Thorndyke and Hayes-Roth, 1979) have found evidence that behaviors associated with these combined memory structures can become enacted without an explicit stimulus. For example, most people can recall an experience of déjà vu—a situation that seems like a memory but you know, rationally, cannot be one. This line of research would explain the phenomenon as the brain enacting an implicit set of memories inappropriately; some cue, a smell, thought, image in your current situation has caused the brain to enact a past memory (template, script, schema, etc.) without all the correct cues (Reber, 1993). If we stipulate that these types of memory optimizers exist, then we may be able to use them in our discussion of intuition and executives.

Enter the works of Herman Ebbinghaus, a mid-nineteenth-century German psychologist, to whom the discovery of the learning curve, the forgetting curve, and the spacing effect are attributed (1885/1913). In his

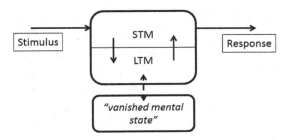

FIGURE 9.2
Generic human memory model.

work, he proposed that mental states of consciousness or experiences are categorized and stored into three categories. In the first category are experiences that can be easily, voluntarily recalled. Using Figure 9.2 as a reference, the stimulus has directly found the response in STM, also appropriately called *active memory*.

In the second category resides those experiences buried a little deeper into LTM and where recall may be more involuntary and need more coaxing; triggered by concentrated thinking or physiologically by a sound, smell, or other association. Again, déjà vu, the feeling of repeating an experience, provides an example most readers can relate to.

Ebbinghaus's third category, the *vanished mental state*, represents the more compelling category for our discussion here. His conjecture (as translated) positions the experiences in this category as hidden from consciousness but still influencing behavior: "Most of these experiences remain concealed from consciousness and yet produce an effect which is significant and which authenticates their previous existence" (1913). Figure 9.2 incorporates Ebbinghaus's level-3 category, the vanished state, to the previous generic human memory model.

INDICATORS OF INTUITION

Experiences that "remain concealed from consciousness" (Ebbinghaus, 1913) sound like our opening dictionary definitions of intuition. Could it be that intuition is the instantiation of Ebbinghaus's third category, the vanished mental state? Could it be that this category can be established through the evolution proposed by Gigerenzer? It seems reasonable that experiences, whether recallable or not, would influence decision making.

The ones we can recall we attribute to experience, but paradoxically, these vanished or lost experiences still influence how we think although we do not consciously know it. Accepting these premises generates follow-on questions. How do we know the *vanished mental state* category exists and how do we see its impact? What does one call these experiences we cannot recall yet have influence on our lives, including decision making?

If the vanished mental state exists and is used by decision makers, then there should be some residual indicators. Much like a wave in a calm lake that can be traced back to a moving boat, we should try to find indicators to trace back that could be attributed to intuition. We propose three such interrelated residual artifacts: *decision-making styles*, *decision style instruments*, and *decision-making strategies*.

Decision-Making Styles

The notion of decision styles has been researched over many years, with multiple perspectives being developed. Decision styles have been examined in the scope of entrepreneurs (Buttner and Gryskiewicz, 1993), hospital administrators (Henderson and Nutt, 1980), and general managers (Pennino, 2002). The problem space or organizational context varies across leadership, strategic planning, negotiations, and so on. There is general acceptance that different cognitive styles prefer and use different sets of information in the decision-making process (Blaylock and Rees, 1984; Nutt, 1986). Rowe and Mason (1987) posit that decision style, while not solely responsible, is a key factor leading to problem-solving success.

The more important question is why we should care about discussing decision styles in a problem-solving and decision-making context. Basically, decision style helps determine what information makes it into the problem space. Given the same data, individuals with differing styles will process the data differently, thereby producing different perspectives on the data—or information.

If the human decision-making process is something like that shown in Figure 9.3, *decision style* matters at every sub-activity in the process. Clearly, the individual's decision style influences the initial perception

FIGURE 9.3
Basic decision making process.

of what constitutes a problem or not. This discovery is slanted based on what the explorer is looking for along with their experiences in similar circumstances. Gathering information, and in fact what information to gather, to understand the problem's scope will be influenced by one's style. We know that some people prefer data displayed in tabular form versus graphical form. The scope of the problem and therefore the areas to look for data will be influenced by decision style and decision strategies. When we put the data back together into a possible solution or solutions, we are synthesizing the data. This synthesis is influenced by the style and strategy of the person or group synthesizing. At the end of the process, when a choice is made from a set of alternatives, that selection process is influenced by the style of the beholder. It is not unheard of to make a decision that conflicts with a committee's recommendation. The implementation of the decision followed by the inevitable interpretation of the decision's success as favorable or unfavorable is slanted further by the decision maker's style and by any observer's style. In summary, the decision style first prejudices the determination if a problem exists. Then it continues by influencing the collection and the interpretation of the data and information along the way, the choice making, and the ultimate response that is made. When this process goes awry for individuals, we call it flawed decision making; when groups of people do this, it takes on the elements of groupthink.

Kahneman et al.'s (1982) work *Judgment under Uncertainty* compiles and categorizes what may be called *anomalies* that individuals project when they solve basic problems. These anomalies or biases are based on the subjects not applying a valid heuristic to obtain the "correct," calculable answer. Interestingly, Kahneman and Tversky alluded to their research in the context of intuition in their postscript to the 1994 printing of their work: "It has been demonstrated that many adults do not have *generally valid intuitions* [emphasis added] corresponding to the law of large numbers, the roles of base rates in Bayesian inference, or the principles of regressive prediction."

Decision Style Instruments

The Decision Style Inventory (DSI) was developed by Alan Rowe et al. (1994) as they tested their conjecture that psychological type was a major element of information systems development and use. The authors have used the DSI to survey and collect data from over 6000 subjects spanning a variety of job levels and occupations. The accumulation of their research is the basis for their book *Managing with Style: Accessing and Improving Decision Making* (Rowe and Mason, 1987). The DSI uses a quick, 20-question, forced-choice questionnaire where each question has four answers that the subject rates exclusively with values of 8, 4, 2, or 1; each rating can be used only once across the four answers to the question. The answers are placed in columns that, when totaled, create a rating for the subject across four decision-making styles: analytical, behavioral, conceptual, and directive. With the amount of data collected over years of research, Rowe and Mason were able to categorize and organize their results with these segments and discuss common styles. The patterns of these segments helped identify the underlying styles.

More broad than the DSI, the Kolbe A™ Index (Kolbe, 2018) attempts to measure one's instinctive way of doing things: *conation*. It makes the distinction between thinking, feeling, and doing. The instrument uses 36 questions to identify natural strengths and compartmentalize these into 12 problem-solving "ways of doing" or Kolbe Action Modes®. Multiple years of research and multiple studies have validated this model and categorization scheme (Kolbe, 2018, 1993). In the end, the instrument reports a method of operation that has influenced job performance and careers.

The Clifton StrengthsFinder® (CSF), another well-researched and well-documented inventory, was developed from the work of Donald O. Clifton (Asplund et al., 2007). The CSF is an online assessment in which each respondent is presented with 177 stimuli and makes 177 responses. Each item lists a pair of potential self-descriptors with five-point Likert anchors from which respondents choose. Most of these descriptors are associated with one of 34 distinct themes of raw talent. The mix of themes then becomes a way to categorize subjects and attribute underlying strengths or talents to these categories.

One of the foundational building blocks of CSF is "Talents are manifested in life experiences characterized by yearnings, rapid learning, satisfactions, and timelessness." The tool was developed to facilitate personal growth and development, but for our discussion, the results of this tool, classifying respondents according to underlying "themes" and talents, seem to represent another tool that matches the search for Ebbinghaus's third category of a vanished mental state.

CSF, Kolbe A, and Rowe and Mason's DSI are not the only decision style instruments that relate to our discussion. The Myers–Briggs Type Indicator (MBTI) test is more popular and has more extensive research and documentation around it. Daniel Goleman's concept of *emotional intelligence* (1995) and the VARK modalities (Fleming and Mills, 1992) also fit well into this discussion. However, the DSI has also been tested and validated extensively and its correlation with the Myers–Briggs test is well documented in *Managing with Style*. The DSI has simplicity on its side and can be administered in a single class period. Returning to the prime definition, this just means that there are many ways to measure decision style.

Decision-Making Strategies

Decision-making models have emerged from many reference areas. The concept of information overload and the mental processes, both the efficiencies and the dysfunctions, created seminal conjectures (Simon, 1955; March and Simon, 1958; Newell and Simon, 1972). For example, *bounded rationality* is one explanation of how individuals organize large and complex sets of information associated with a decision. A second, the satisficing strategy, was derived as a means to explain how individuals minimize the impact of that information. A satisficing strategy can also explain why individuals could choose non-optimal decisions wherein locally optimal solutions win out over globally optimal answers. For example, a shopper parking his or her car may choose to take the first parking spot available or to troll the parking lot looking for a better spot (i.e., one closer to the store). The decision in the first case is satisficing in that it was the first parking spot available that met the shopper's criteria. In a more detailed example, Gigerenzer (2007, p. 56) uses this strategy to relate his fable of robot love and how brains evolve.

Similarly, Butler and Loomes's (1997) aspiration-satisficing model was developed in an economic context where subjects searched for the best price. This model extends the base satisficing model to include an adjustment to the individual's aspiration level, after each price quote is received, by incorporating the following rule:

> Set some initial aspiration level and begin to search. After each quote, modify aspirations to take account of the quote(s) obtained so far. Continue searching until you have received a quote that allows you to do as well or better than your current aspiration level. (p. 133)

John D. Hey's work (1982, 1986, 1994), again in the arena of obtaining price quotes, extracted additional strategies. Hey developed a series of processing rules to help explain the differences found between the optimal decision strategy and those exhibited by his subjects in his designed research. His reservation rule and its derivative, the optimal reservation rule, also work to incorporate the satisficing idea over multiple periods of decision making. His bounce strategy, wherein the rule says to stop if the current quote is less than the first quote, acknowledges that decision making incorporates an order effect. The comparison of the current quote with the previous quote, as dictated by the rules, provides information to the individual for their current decision process.

The works of Kahneman and Tversky (1982), Kahneman et al. (1991), Knetsch (1989), and Thaler (1980) look at the decision maker and his or her reference point. This research argues that decisions made by individuals differ based on their point of reference to the choices offered. Extending from this reference-dependent model, the loss aversion decision strategy proposes that individuals perceive an incremental loss different from the same absolute incremental gain (Knetsch, 1989). From there, the basic prediction, supported by the research, is that subjects will choose to minimize their potential loss versus the potential gain when, rationally and quantitatively, there is no difference. A situation where there is a potential $250 gain is chosen less often than a situation with a potential $250 loss. Another derivation by Samuelson and Zechhauser (1988), the status quo bias, explains why decision makers may be reluctant to make a change decision.

Ellen Langer (1975) offers a slightly different explanation. She suggests that decision makers may make less than optimal decisions because the decision maker feels that he or she has an unrealistic impact on the results. She named this phenomenon the *illusion of control* and defined it as "an expectancy of a personal success probability inappropriately higher than the objective probability would warrant." There are real-world examples of this phenomenon. Mason and Mitroff (1974) did an extensive study of scientists and their interpretation of the information provided by the moon rocks brought back from the Apollo moon missions. Detailed interviews were used to identify changes or impacts on belief structures held by scientists as they evaluated the moon rocks. Ultimately, the study documented the role played by what were termed *irrational* factors in the scientists' resulting belief structures.

In addition to Langer's illusion of control model, other decision-making models seem to represent indicators of some hidden underlying process in humans contributing to decisions. The satisficing and the reference-dependent model are two such models. Martz et al. (2006) used a fictional venture capital situation to look at decision making and these models along with the illusion of control. Here the subjects—senior business students—were placed in a hypothetical managerial setting for a venture capital firm where they had to make a decision concerning a limited resource and how to apply that resource based on each investment in 10 companies. If one believes that the case was unique and not known to the students, then their decision-making patterns could be seen as applying contributions from their LTM and any *vanished mental state,* intuition to solving the problem. The results pointed to strategies that were reference dependent and adaptive and demonstrated a decision-making pattern that the *illusion of control* could explain.

In summary, decision-making studies continue to show the interaction of decision-making styles, instruments, and strategies. It is clear that an immense number of factors impact, supporting or inhibiting, an individual as he or she makes a decision. Researchers such as Kahneman and Tversky (1982), Langer (1975), Simon (1955), March and Simon (1958), and Mason and Mitroff (1974) have contributed to this body of research and defined several working strategies concerning human decision making.

CONCLUSION AND TAKEAWAYS

Combining the elements discussed in this chapter, we can visualize the impact of intuition, through decision style, on the basic problem-solving process (Figure 9.4).

If we believe intuition impacts decision making, we must look for possible indicators of this influencer. There is a body of research that proposes decision styles may be acting according to our definition of intuition. Decision-making styles sit as one possible example of the vanished mental state of decision makers' minds that implicitly—without reaching

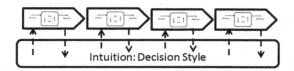

FIGURE 9.4
Intuition's impact on decision making.

the active STM—influences decision making. Over time, behaviors operating as decision-making strategies have been discovered, named, and associate with styles. The point of this chapter is that decision styles and their associated strategies look like possible examples of intuition, as they influence the whole decision-making process from beginning to end without the explicit knowledge of the decision maker.

There are several takeaways for readers. The first is that cognitive scientists are making headway with understanding better how the human brain stores and retrieves information. Second and concurrently, there is growing acknowledgment that there is more involved with decision making than the traditional rationalistic method. Third, decision style instruments are exposing what seem to be decision styles and decision-making strategies that may help define a type of intuition. All of these takeaways combine to provide the beginning of a roadmap for a better understanding of how intuition and decision making interact by means of decision styles and decision strategies.

REFERENCES

Asplund, J., S.J. Lopez, T. Hodges, and J. Harter (2007). *The Clifton StrengthsFinder® 2.0. Technical Report: Development and Validation*, The Gallup Organization, February 2007. http://menedzsmentor.com/customer_documents/cikkek/csftechnicalreport 031005.pdf. (last accessed May 19, 2018).

Atkinson, R.C., and R.M. Shiffrin. (1968). Human memory: A proposed system and its control processes. In K.W. Spence and J.T Spence, *The Psychology of Learning and Motivation*, pp. 89–195. New York: Academic Press.

Bartlett, F.C. (1932). *Remembering: A Study in Experimental and Social Psychology*. London: Cambridge Press.

Blaylock, B.K., and L.P. Rees. (1984). Cognitive style and the usefulness of information. *Decision Sciences*, Vol. 15, pp. 74–91.

Butler, D., and G. Loomes. (1997). Quasi-rational search under incomplete information. *The Manchester School of Economic and Social Studies*, Vol. 65, No. 2, pp. 127–144.

Buttner, E.H., and Gryskiewicz, N. (1993). Entrepreneurs' problem solving styles: An empirical study using the Kirton adaptation/innovation theory. *Journal of Small Business Management*, Vol. 12, No. 1, pp. 22–31.

Cambridge Dictionary. https://dictionary.cambridge.org/us/dictionary/english/intuition (last accessed May 19, 2018).

Cherniak, C. (1984). Prototypicality and deductive reasoning. *Journal of Verbal Learning and Verbal Behavior*, Vol. 23, pp. 625–642.

Ebbinghaus, H. (1885). Translated by Henry A. Ruger and Clara E. Bussenius (1913). *Memory: A Contribution to Experimental Psychology*. New York, NY: Columbia University Press.

Feigenbaum, E.A., and H.A. Simon. (1984). EPAM-like models of recognition and learning. *Cognitive Science*, Vol. 8, pp. 305–336.

Fleming, N.D., and C. Mills. (1992). Not another inventory, rather a catalyst for reflection. *To Improve the Academy*, Vol. 11, p. 137.

Gigerenzer, G. (2007). *Gut Feelings: The Intelligence of the Unconscious*. New York: Viking.

Goleman, D. (1995). *Emotional Intelligence*. New York: Bantam Dell.

Greenspoon, J. (1955). The reinforcing effects of two spoken sounds on the frequency of two responses. *American Journal of Psychology*, Vol. 68, pp. 409–416.

Henderson, J., and Nutt, P. (1980). The influence of decision style on decision making behavior. *Management Science*, Vol. 26, No. 4, pp. 371–386.

Hey, J.D. (1982). Search for rules for search. *Journal of Economic Behavior and Organization*, Vol. 3, No. 1, pp. 65–81.

Hey, J.D. (1986). Still searching. *Journal of Economic Behavior and Organization*, Vol. 8, pp. 137–144.

Hey, J.D. (1994). Expectation formation: Rational or adaptive or ...? *Journal of Economic Behavior and Organization*, Vol. 15, pp. 329–349.

Jenkins, J.G. (1933). Instruction as a factor in incidental learning. *American Journal of Psychology*, Vol. 45, pp. 471–477.

Kahneman, D., P. Slovic, and A. Tversky (Eds.) (1982). *Judgment under Uncertainty: Heuristics and Biases*. Cambridge, MA: Cambridge Press.

Kahneman, D., J. Knetsch, and R. Thaler. (1991). Anomalies: The endowment effect, loss aversion, and status quo bias. *Journal of Economic Perspectives*, Vol. 5, No. 1, pp. 193–206.

Knetsch, J.L. (1989). The endowment effect and evidence of nonreversible indifference curves. *The American Economic Review*, Vol. 79, No. 5, pp. 1277–1284.

Kolbe Corp. (2018). *Research and Validation*. http://kolbe.com/why-kolbe/history-and-expertise/research-and-validation (last accessed May 18, 2018).

Kolbe, K. (1993). *Pure Instinct*. New York: Times Books.

Langer, E. (1975). The illusion of control. *The Journal of Personality and Social Psychology*, Vol. 32, pp. 311–328.

March, J.G., and H.A. Simon. (1958). *Organizations*. Cambridge, MA: Blackwell.

Martz, W.B., Jr., and M.M. Shepherd. (2003). *Testing for the Transfer of Tacit Knowledge*. *Decision Sciences Journal of Innovative Education*,. Vol. 1, No. 1, Spring, pp. 41–56.

Martz, W.B., T. Neil, and A. Biscaccianti. (2006). Exploring entrepreneurial decision-making strategies. *International Journal of Innovation and Learning*, Vol. 3, No. 6, pp. 658–672.

Mason, R.O., and I. Mitroff. (1974). On evaluating the scientific contribution of the Apollo moon missions via information theory: A study of the scientist–scientist relationship. *Management Science*, August, Vol. 20, No. 12, pp. 1501–1575.

Merriam-Webster's Dictionary. https://www.merriam-webster.com/dictionary/innate (last accessed May 19, 2018*).*

Newell, A., and H. A. Simon. (1972). *Human Problem Solving*. Englewood Cliffs, NJ: Prentice-Hall.

Nutt, P.C. (1986). Types of organizational decision processes. *Administrative Science Quarterly*, Vol. 29, No. 3, pp. 414–450.

Osborn, A.F. (1963). *Applied Imagination*, New York: Scribner.

Papert, S. (1980). *MindStorms*. New York, NY: Basic Books.

Pennino, C.M. (2002). Is decision style related to moral development among managers in the U.S.? *Journal of Business Ethics*, Vol. 41, No. 4, pp. 337–347.

Piaget, J. (1929). *The Child's Conception of the World*. New York: Harcourt, Brace.

Polya, G. (1957). *How to Solve It*, 2nd ed. Princeton, NJ: Princeton University Press.

Reber, A.S. (1993). *Implicit Learning and Tacit Knowledge*, Oxford Psychological Series Vol. 19.

Roby, T.B. (1966). Self-enacting response sequences. *Psychological Reports*, Vol. 19, pp. 19–31.

Rowe, A.J., and R.O. Mason. (1987). *Managing with Style*. New York: Wiley.

Rowe, A.J., R.O. Mason, K.E. Dickel, R.B. Mann, and R.J. Mockler. (1994). *Strategic Management: A Methodological Approach*. Reading, MA: Addison-Wesley.

Saaty, T. L. (2000). *The Brain: Unraveling the Mystery of How It Works*. Pittsburgh, PA: RWS.

Samuelson, W., and R. Zeckhauser. (1988). Status quo bias in decision making. *Journal of Risk and Uncertainty*, Vol. 1, pp. 7–59.

Sandelands, L.E., S.J. Ashford, and J.E. Dutton. (1983). *MotivEmot*, Vol. 7, p. 229. https://doi.org/10.1007/BF00991675.

Schank, R., and R. Abelson. (1977). *Scripts, Plans, Goals and Understanding*. Hillsdale, NJ: Lawrence Erlbaum Associates.

Shannon, C.E., and W. Weaver. (1949). *The Mathematical Theory of Communication*. Urbana: University of Illinois Press.

Simon, H. (1955). A behavioral model of rational choice. *The Quarterly Journal of Economics*, Vol. 69, pp. 99–118.

Thaler, R. (1980). Toward a positive theory of consumer choice. *Journal of Economic Behavior and Organization*, Vol. 39, pp. 36–90.

Thorndike, E.L., and R.T. Rock, Jr. (1934). Learning without awareness of what is being learned or intent to learn it. *Journal of Experimental Psychology*, Vol. 17, pp. 1–19.

Thorndyke, P., and B. Hayes-Roth. (1979). The use of schemata in the acquisition and transfer of knowledge. *Cognitive Psychology*, Vol. 11, pp. 82–106.

Warfield, J.N. (1989). *Societal Systems*. Seaside, CA: Intersystems.

Warfield, J. (1990). *A Science of Generic Design*. Salinas, CA: Interscience.

Whiting, C.S. (1958). *Creative Thinking*. New York, NY: Reinhold Publishing.

Zander, T., M. Ollinger, and K. Volz. (2016). Intuition and insight: Two processes that build on each other or fundamentally differ? *Frontiers in Psychology*, Vol. 7, pp. 1395.

10

Improvisation and Instinct

Written by: William Hall and Rebecca Stockley

CONTENTS

There are people who prefer to say 'yes' and there are people who prefer to say 'no'.. Those who say 'yes' are rewarded by the adventures they have. Those who say 'no' are rewarded by the safety they attain.

—Keith Johnstone

INTRODUCTION

When you hear the word *improvisation* it may bring to mind a jazz group, or dancers, or a theater company. It is most often used to describe a type of performance: *The art or act of improvising, or of composing, uttering, executing, or arranging anything without previous preparation.* What do all these have in common to qualify them as improvisation? The unique quality of all improvisation is that the creative act is not pre-planned; it happens in the moment. It is created and displayed in the moment. How can a group create something without previous preparation and how can they expect it to be cohesive and worthy of accolades?

For many years, we have been using improvisational skill sets and principles in order to provide a new way to connect with others and to respond quickly with spontaneity. These skills, often referred to as *soft skills*, are the skills of interaction that build connections, release creativity, and invite authenticity: in other words, the skills of leadership. It's all done with lessons in improvisational theater. Improvisation may also lead to improving one's intuitive awareness.

Our goal is not to turn you into an improv performer but instead to increase your ability to be a spontaneous collaborator able to work in real time. But before we can do that, we have to confront the biggest obstacle: fear.

One interesting thing that happens when we show up in your workplace to lead a workshop in applied improvisation is we meet frightened participants. "Ack! You're the improv people! I'm really nervous!" When people learn that they are taking an improvisation workshop, their heartbeat increases, their pupils dilate, and the fight or flight instinct kicks in.

We have instincts to keep us safe. This instinct can prevent us from going into the unknown. The instinct to be safe can urge us to resist change. It can keep us back on our heels, tacitly watching the world go by, rather than leaning into it and engaging with it.

> *One of the secrets to staying young is to always do things you don't know how to do, to keep learning.*
>
> **—Ruth Reichl**

For some people, the idea of taking an improvisation workshop can trigger fear. We think of improv as fast, clever, creative. The idea that you will be expected to be creative on demand can trigger fear, and why not? If our participants are overwhelmed with fear, creativity is out of reach. Once fight or flight is triggered, it's difficult to learn. With fearful participants, our approach is to gradually enter a world where fear is unnecessary as well as detrimental to the outcome. This approach is the same as working with people who fear public speaking. Rather than starting with a TED Talk, start with a toast at a friend's wedding.

Line of Fear: The personal barrier separating comfortable activities and behaviors from uncomfortable, unknown or frightening activities.

You can waste your lives drawing lines. Or you can live your life crossing.

—**Shonda Rhimes**

Through improvisation classes, people can learn to move that line of fear of public speaking, giving them more choices in critical situations. In the practice of improvisation, we demystify the creative process and move the line of fear about being creative on demand. We use improvisation to explore spontaneity and engage the imagination, moving the line of fear of thinking on your feet. Whether it's improv, or leading a new team through a new project, or jumping out of an airplane, the line of fear can be moved.

First, we play. In our improvisation workshops, we engage participants in different kinds of playful activities. Play relaxes people. Play keeps people in the moment. Play is fun. Play creates harmony in a group. While it may only be for the duration of the game that you're playing, play can bring people together. Play can lead us into taking risks without fear.

Play touches and stimulates vitality, awakening the whole person—mind, body, intelligence and creativity.

—**Viola Spolin**

The business world can be very competitive, a battlefield where there are winners and losers, a zero-sum game. The creative process, the improvisational frame, is not competition, it's collaboration. In fact, it's more than collaborative; we believe we spark the best out of each other. We use this phrase in improvisation to capture the idea: "Make your partner look good." The benefit of working to make your partner look good is engagement.

Second, we engage. We encourage you to work with your partner, rather than just tolerating him or her. Engage with your partner. Listen to what he or she says, get curious about his or her ideas. Let your team members shift your perspective and it will influence your experience. Let yourself be changed by your partner.

We are going to share with you some methodologies for moving your line of fear. Improvisation will not remove your instincts to stay safe; those instincts will still be there to protect you. This is not about ignoring the great guidance fear can provide, it's about redefining the unknown for yourself so you can step onto the stage of your life fearlessly and confidently, engaging your family, your community, and your team.

Improvisation is an experiential skill. Throughout this chapter, we will be sharing exercises you can do either by yourself or with your colleagues. We encourage you to bring both curiosity and courage to the experiences.

FEAR

You go on ahead and investigate the noise outside in the dark, I'll stay in the cabin.

—Horror movie cliché

We have a primal instinct to survive; that instinct has shaped the evolution of humans to be first fearful and relaxed later. That instinct can show up in our lives when we are walking across a field and we suddenly find ourselves jumping away from a stick in the grass. Our brains didn't engage the prefrontal cortex and do a comparison of snakes versus sticks. Because if it had been a snake, that evaluation would have taken too long. We might have been poisoned. The fear reflex is faster than conscious thought: You are dicing carrots in the kitchen, your eyes snap shut, and a tiny piece of carrot hits your eyelid. Before you are consciously aware of something shooting toward your eye, your instinct protects you. With this reflex comes a very healthy mindset of being surrounded by danger. Fear can work successfully to tell us we're in danger without conscious awareness. Fear can also be triggered in situations that are not dangerous. Stage fright, performance anxiety, and flop sweat are vestigial fear reactions that improvisation training can help reframe.

You can't be that kid standing at the top of the waterslide, overthinking it. You have to go down the chute.

—Tina Fey

Our distant ancestors faced terrible and very real fear. We're hardwired not to step into danger. Imagine you're in the woods seeking shelter from a storm. You find a cave and consider going in. As you listen from the cave entrance, you can hear growling and roaring from inside. Do you go into the cave? Fear says, "No!" But improvisation training says, "Yes!" In improvisation, where there is no actual danger, that instinct can prevent adventure and story.

Go into the cave!

—Keith Johnstone

When we are faced with a situation that is uncomfortable and stressful we naturally default to avoidance. Combat troops go through special training to be able to face fearful situations by giving them experience and building skills through practice. Improvisation classes train the body and brain to acknowledge fear and reframe it as excitement, a signal of the unknown. In improvisation, the unknown is an opportunity.

When we are relaxed and grounded, we see everything. Our senses are sharp and responsive. But as fear creeps in, our vision narrows until we no longer see anything or anybody.... We even have difficulty seeing ourselves.

—**Shawn Kinley**

When we play an improvisation game, we are engaging in a kind of make-believe that in practical terms eliminates any real danger.

Try this: Activate Your Imagination

Take a moment and bring to your mind a good meal you had recently. Imagine where you were. Who was with you. The anticipation when you saw your meal listed on the menu. The presentation when it arrived at the table. And finally, the taste as you brought it to your lips.

Did your mouth water? If so, your vivid imagination was activated. The mind has difficulty separating reality from imagination. Even when we create a 100% fictitious situation, the instincts that keep us safe in the 100% real world are still at work.

Through improvisation workshops, each participant moves his or her own lines of fear. This can make the unknown approachable, and going into the unknown can be life changing.

PLAY

Play isn't the enemy of learning, it is learning's partner

—**Stuart Brown**

When people gather for their first improv workshop, we intentionally start with activities that look and feel familiar. Many adults at work spend time in conversation. Starting with a conversation game feels easy. Often called Icebreakers, these games connect people and help them relax.

Try This: Three Things in Common in Three 3 Rounds

Round 1: Find a partner. Use conversation to discover three things you have in common with your partner—things you're not aware of before you begin.

When we lead this activity, we give a 2-minute time limit for the discovery of three commonalities. As people play, we notice the body language becomes looser, people relax, and the room fills with social laughter. After 2 minutes, we stop the game and ask people to reveal their discoveries.

It can be amazing the connections that people find. In one workshop, two women learned that they had attended the same girls' high school, in the same town in India, but 20 years apart!

Round 2: With the same partner, find three things in common a bit deeper than during the first round.

If someone asks what we mean by *deeper*, it can be a delaying tactic to prevent going into the unknown. We playfully reply, "Decide between yourselves," and start the timer. As the conversation flows, we can hear that the vocal tones of the participants are a bit more thoughtful. The people in the room seem more attentive and calm.

After two minutes, we ask people to reveal some of their commonalities.

Round 3: Find 3 three things in common without using language. The facilitator often introduces this round without speaking himself—using only mime and charades skills. This will often lift the mood higher and remind us that we are communicating with our bodies and facial expressions. Everything we do is a communication. The reveal of commonalities after the non-verbal round can be without language as well.

In the Three Things in Common activity, three things have happened to make the participants more comfortable, to move them away from fear. First, when we play, the focus is no longer on the individual but on the partner. Second, we begin to see each other as richer human beings.

Third, familiarity reduces our anxiety as we discover commonalities with others.

When we can reduce fear and increase trust, a richer set of possibilities appears. We are more grounded and more interested in the process and in relationships than we are when we act from a place of anxiety and fear. Anxiety leads to fear and then aggression and then hostility for some. For others, anxiety leads to fear and then disengagement and isolation. As leaders, we walk a narrow path in developing a team that is relaxed, trusting, and open to greater possibilities.

As improvisation instructors, we make the experience safe by reminding participants that they're already experts in improvisation. We don't have a script for our lives. Every interaction is an improvisation. Frankly, while this may appeal to the mind, it does very little to reduce fear.

No one knows what's next, but everybody does it.

—George Carlin

In business, accuracy is vitally important, so we hire individuals who get things right. But if you're a perfectionist, uncomfortable making mistakes, it can make creativity inaccessible. A healthy relationship with acceptable risk and mistakes is invaluable to the processes of innovation, creativity, and transformation.

We use a variation of this game to redefine our relationship to failure. Two by Three by Bradford created by Augusto Boal:

Find one person with whom to play this game. Face each other and count from one to three, alternating between you. A says "One," B says "Two," A says "Three," and repeat:, B says, "One," A says "Two," and continue ad infinitum.

Now that you've played *Two by Three by Bradford*, you may have noticed that you used quite a bit of effort to "get it right." Your focus and attention on your partner and the sequence became acute. Your breath may have become shallow. You may have struggled with it, become serious, laughed. Some people experience embarrassment or frustration when playing this game for the first time. We all respond to this deceptively simple challenge in different ways; but more often than not, people make mistakes when learning to play this game. When we make mistakes where we assume we will not, our failure response takes us out of play. Your response may be to laugh or to furrow your brow and try harder, but whatever that response is, it's isolating you from your partner rather than connecting you to him or her. What would happen if we flipped the response? What if you celebrated your failures together with a pose of triumph?

When people win, achieve a victory, or celebrate success, there is a cultural universal gesture: the pose of triumph.

> *Try this: The Failure Bow. Stand up, put your feet at least shoulder- width apart with your weight balanced equally over both feet. Make your spine long, lift your chin, throw your arms way up in the air, take a breath, and cheer with joy. Check your arms, are they fully extended over your shoulders? If not, try it again with your elbows up above your shoulders. Chin up, big breath, cheer with joy.*

Human beings celebrate big wins with this action. In improv training, we call this the Failure Bow. Many practice the Failure Bow with a phrase like, "I made a mistake!"

What if you did this when you tried something new and failed? You take a risk, like playing a new game, you fail, celebrate, and then you start to play again. In practice, this celebration of mistakes can shift your relationship to mistakes, thereby shifting your relationship to risk and creating more opportunities for innovation. In the words of UCLA basketball coach John Wooden, "*If you're not making mistakes, you're not doing anything. I'm positive that a doer makes mistakes.*"

Play *Two by Three by Bradford* again, and this time, celebrate when you make a mistake. Make your partner look good by celebrating when he or she makes a mistake. *Make the game more challenging by replacing "One" with a clap of your hands. Then replace "Two" with a snap of your fingers. When it's your turn to say "Three",," hop instead.* Play the game and continue to celebrate when you make mistakes. In so doing, you will get better at the game, and you'll have fun doing it. By celebrating acceptable risk and failure, you may learn to take risks, fail, and stay happy and engaged.

Start with yourself and reprogram your relationship to failure. If a team is made up of individuals who are engaged and willing to take risks, the team will have a positive, dynamic balance.

As mentioned before, we teach improvisers to retrain restrictive, controlling, and isolating instincts that fear can inspire. We do this by using "yes."

Yes. The word "yes" and the concept of "yes."

Bad improvisers block action, often with a high degree of skill. Good improvisers develop action.

—Malcolm Gladwell

In 1967, in central London, a young musician went into a tiny gallery to see an avant-garde art exhibition. Inside the door there was a stepladder. On the ceiling above the ladder was a spyglass hanging from a chain. Attached to the ceiling was a message painted in tiny letters. The musician recalls being interested in discovering what was painted on the ceiling. He said he found it a bit foolish and dangerous to climb the ladder and read the word. He thought it was going to say something disparaging. But when he finally got to the top, he looked through the spyglass and saw a word that changed his life forever. The word was "yes." The musician had to follow the "yes" and meet the artist. That is how the word "yes" brought John Lennon and Yoko Ono together.

"Yes" keeps things moving and "no" stops them. The secret to everything that has ever been done in civilization is saying "yes" to action.

Try This: Shared Adventure

Play this game with one partner. First, decide on a location in the world where you would both like to visit. Now, imagine you are good friends who have not seen each other for a while. Imagine that you went on an adventure together in that place years ago. It was great. One partner starts with, "Remember our adventure in [name place]? The other person has a choice. The partner can say, "No"," or they can say "Yes".." The first speaker tries again:, "Remember when we hiked to the volcano?"." Reply with "Yes" or "No" again.

Try co-creating the story of your adventure using "yes" or "no" in response to your partner for a minute.

How did that go? You may have noticed it was challenging to construct the story of your adventure in this conversation.

The "no" stops the conversation. If your partner stops you with a "no," you can try again; suggest other aspects of your vacation. If you're alternating "yes" and "no" in response, the possibilities may open up briefly, but the conversation doesn't flow or get anywhere.

"Remember our adventure in Hawaii?"
"No."
"Remember? We stayed in that bungalow on the beach?"
"Yes."
"Remember we took that surfing class?"
"No."

You can try to spark the conversation, but when we're faced with a partner who replies, "No," we exhaust our ideas and the shared adventure doesn't come to life.

Now add the secret word "and".." The first person says, "Remember our
   adventure in [name of place].
The next speaker begins with, "Yes, And..."
You have a new tool: "Yes, And..."
You: "Remember our trip to Hawaii?"
Your partner: "Yes, and we had the best time surfing."
You: "Yes, and we found a secret beach with the best waves."
Your partner: "Yes, and the dolphins would swim with us as we were
   surfing."
You: "Yes, and..."

You may notice that sometimes the "Yes, and..." flows easily from idea to idea. You may pick up momentum and surprise even yourself. Sometimes your ideas may not flow with ease. If your partner's idea doesn't inspire you, you can add more detail to the previous idea and the flow will pick up again.

Saying "no" is a great leadership tool for many occasions but not all. If you say "no" because you are fearful or feel out of your comfort zone, it will close down the people you lead, eventually lowering their engagement. This can prevent the co-creation of wonderful possibilities.

Notice when you or members on your team say "no." And then ask yourself if the reason might be fear. It takes courage to say "yes." We never know what will come next. The rewards are increased engagement and a leadership style grounded in moving forward.

A word of warning: We have taught this concept to many companies around the world, and occasionally it is used as a weapon. In a meeting, a team is being asked to do more work without additional resources. One of the team members says, "We can't meet the timeline without additional people on the team." His manager replies, "Yes we can. Remember our improv workshop? We have to say, 'Yes, and...'" This behavior isn't useful. Saying "Yes, and..." is a very useful technique or tool for moving things forward, but it is not a rule for enforcement. There is a place for "yes." Whenever you're creating something new, inventing, innovating, brainstorming, or kicking ideas around, "Yes, and..." is helpful. It isn't very useful when you're negotiating, evaluating, selecting, or making decisions.

ENGAGE

Control leads to compliance; autonomy leads to engagement.

—Daniel Pink

How Can We Create A Safe Space For All Your Colleagues And Employees?

When we improvise, we share the stage. Sometimes there are 10 people on stage. If all of us are speaking at once, it's chaos. When training improvisers, we work on teaching everyone to speak up and follow their impulses; but to be honest, the impulse to speak up comes more easily for some than for others. It's common to see improv performances where one or two of the people onstage do most of the talking while others remain in the background. We call this *driving*. In the process of building an ensemble, we've learned to teach drivers to listen more deeply and to teach those who are quieter to speak up more. The give and take of focus on the stage is a parallel to the give and take of control on a work team.

We have a tremendous diversity of people in the workplace today. One difference we've encountered is related to social dynamics. Some people seem socially comfortable, some seem more socially uncomfortable. These behaviors can be attributed to many things: experience, expertise, age, culture, respect, gender, and education. These things and more can determine communication style in group situations. When you have a meeting, it is likely that some of your team will speak up, contribute ideas, and even fight for them, while others rarely speak up.

How might we make it safe for the quiet people to contribute? Teach *Deep Listening* to your team and you can create the space for everyone to speak up. There is a great exercise for the practice of deep listening, *Zen Counting.*

Try This: Zen Counting

Gather a group of 5 to 16 people in a circle. Give the following instruction: We're going to count to 20 in the usual order. Anyone can say, "One".. Someone else will say, "Two".. Then, "Three".. We'll go on in the usual ascending numeric order until we get to 20. No one person says two numbers in a row. If two or more voices says the same number, we go back to "One".. Use listening as your strategy. Count spontaneously, no one directs who speaks next. Let's see how high we can get.

Then try it. You may find that you cannot get beyond 3 or 4. Try again! You may get to 20 on the first try. Well done! Celebrate! Then try to get to 30.

In Zen Counting, you may notice an emergent strategy: Perhaps people were speaking in order around the circle, or an individual "claimed" certain numbers, or people could be rushing to count. If you notice a strategy, identify it and ask people to focus on listening.

We've had the experience where we introduce this activity to a group of people and they can't get to 10. In another situation, a group of people got to 100 the first time we tried it. What's the difference? Deep listening! What's our coaching for participants? "Breathe deeply, listen, and try again."

Respect helps us listen deeply to our fellow humans. Therefore, when you listen deeply to someone, he or she feels respected.

How Might We Create a Safe Space for The Fearless "Yes, and..."?

The stage is a safe place for imagination. The things we do and say on stage do not necessarily represent who we are or what we believe. In Shakespeare's *Richard III*, the title character is written as an evil man. The actor playing the role doesn't take the responsibility or blame for the character's behavior; no actor would want to play Richard III if when he came offstage he was subjected to personal harsh judgment. "How dare you kill your father and seduce your stepmother?!"

What is the equivalent space for a leader? Make the wrong decision and you may be out of a job or the company could be destroyed. The answer is simpler in concept than it is in action: Defer your decisions. This deferral will allow you freedom to explore without consequence a limitless world of possibilities and ways to move forward. Make space for ideas to flow. Consider the ideas. Then make the decision. To do this you need to separate the generation of ideas from their evaluation. Make the process transparent, clarify that no decisions are being made, no projects funded, no reorganization scheduled during the brainstorming process. Create a "Yes, and..." space.

When we have space for exploration, we place a higher value on sparking each other's interest than on getting to a solution. We value contribution above feasibility. Reassure yourself that the evaluation and decision will come later in the process. Evaluating ideas as they appear dampens the creative energy, slows the pace of contribution, and shuts people down. Freewheeling creativity cannot happen simultaneously with evaluation. They are opposites.

The best plans often start out as wild ideas.

—**Randy Nelson**
Former Dean of Pixar University

What Is In a Wild Idea?

Wild ideas are big ideas, crazy ideas, and quite possibly stupid ideas. But within a wild idea may lay the seed of a breakthrough idea. The world is full of examples, from a couple of guys building computer boards in a Los Altos garage creating Apple, Inc., to a couple of other guys who could not afford the rent on their San Francisco apartment creating Airbnb.

TAKE AWAY

What Will You Do Differently?

Move your line of fear. Redefine your relationship to failure. Create opportunities for your team to play. Create a safe zone for your team to say "yes" to all ideas. Get everyone engaged. Move the line of fear for your team, your division, your company, and lead them into the unknown where the possibilities are infinite. These helpful hints should improve your intuitive awareness.

Index